Butterflies

of the Caribbean
and Florida

Butterflies

of the Caribbean and Florida

PETER STILING

CARIBBEAN

First published 1999 by
MACMILLAN EDUCATION LTD
London and Basingstoke
Companies and representatives throughout the world

ISBN 0–333–73573–0

10   9   8   7   6   5   4   3   2   1
08   07   06   05   04   03   02   01   00   99

This book is printed on paper suitable for recycling and
made from fully managed and sustained forest sources.

Typeset by EXPO Holdings, Malaysia

Printed in Hong Kong

A catalogue record for this book is available from the
British Library.

Front cover illustration: Postman (*Heleconius melpomene*)
Back cover illustration: Silverking (*Archaeoprepona demophoon*)

Acknowledgements

All the photographs are by the author except for the following:

The Queen	Leroy Simon	Visuals Unlimited
Wood Nymph	Bill Beatty	Visuals Unlimited
Ruddy Daggerwing	Leroy Simon	Visuals Unlimited
Ruddy Daggerwing caterpillar	Leroy Simon	Visuals Unlimited
Ruddy Daggerwing pupa	Leroy Simon	Visuals Unlimited
Mosaic caterpillar	Leroy Simon	Visuals Unlimited
Cracker	Leroy Simon	Visuals Unlimited
Viceroy	John Gerlach	Visuals Unlimited
Viceroy caterpillar	James L. Castner	
Painted Lady	Leroy Simon	Visuals Unlimited
Grey Hairstreak	Leroy Simon	Visuals Unlimited
Cassius Blue	Leroy Simon	Visuals Unlimited
Florida White	Ken Brate	Photo Researchers Inc.
Little Sulfur	Ken Brate	Photo Researchers Inc.
Dwarf Yellow	Leroy Simon	Visuals Unlimited
Orange Sulfur	Bill Beatty	Visuals Unlimited
Dogface	Leroy Simon	Visuals Unlimited
Cloudless Sulfur caterpillar	Leroy Simon	Visuals Unlimited
Zebra Swallowtail caterpillar	James L. Castner	
Green-clouded Swallowtail caterpillar	James L. Castner	
Palamedes Swallowtail caterpillar	James L. Castner	
Androgeus Swallowtail-male	Leroy Simon	Visuals Unlimited
Androgeus Swallowtail-female	Leroy Simon	Visuals Unlimited
Androgeus Swallowtail caterpillar	Leroy Simon	Visuals Unlimited
Long-tailed Skipper caterpillar	Leroy Simon	Visuals Unlimited
Sickle-winged Skipper	Ann B. Swengel	Visuals Unlimited

The author gratefully acknowledges the help of Cathy Ringrose in locating these slides.

The author and publishers would like to thank the Desk Gallery, Zedcor Inc. for permission to use the butterfly images in the tables.

Contents

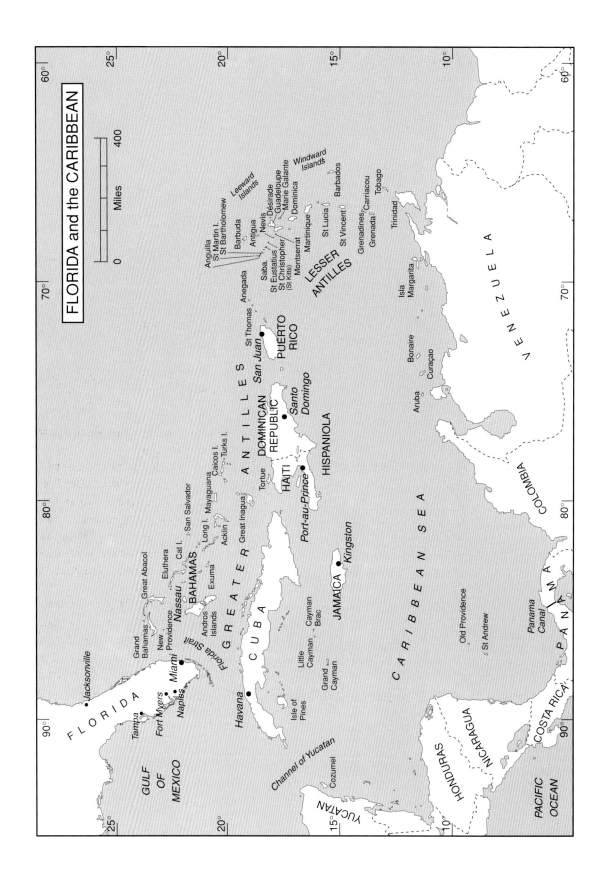

FLORIDA and the CARIBBEAN

Miles

0   400

25°

20°

15°

10°

60°

70°

80°

90°

FLORIDA

Jacksonville

Tampa

Fort Myers

Naples

GULF
OF
MEXICO

Grand
Bahamas

Great Abaco

New
Providence

Nassau

Andros
Islands

BAHAMAS

Eleuthera

Cat I.

Exuma

San Salvador

Long I.

Acklin

Mayaguana

Caicos I.

Turks I.

Great Inagua

GREATER   ANTILLES

Miami

Florida Strait

Havana

CUBA

Isle of
Pines

Little
Cayman

Cayman
Brac

Grand
Cayman

JAMAICA

Kingston

CARIBBEAN   SEA

Old Providence

St Andrew

Channel of Yucatan

Cozumel

YUCATAN

HONDURAS

NICARAGUA

COSTA RICA

PANAMA

Panama
Canal

PACIFIC
OCEAN

Tortue

HAITI

Port-au-Prince

DOMINICAN
REPUBLIC

Santo
Domingo

HISPANIOLA

San Juan

St Thomas

PUERTO
RICO

Anegada

Anguilla

St Martin I.

St Bartholomew

Barbuda

Antigua

Nevis

Saba

St Eustatius

St Christopher
(St Kitts)

Montserrat

Leeward
Islands

Désirade

Guadeloupe

Marie Galante

Dominica

Windward
Islands

Martinique

St Lucia

St Vincent

Grenadines

Grenada

Carriacou

Tobago

Barbados

Trinidad

LESSER
ANTILLES

Isla
Margarita

Bonaire

Curaçao

Aruba

V E N E Z U E L A

COLOMBIA

Introduction

This book illustrates the most common and interesting butterfly species of the Caribbean and Florida. The selection of the species has been guided by the intention to present photographs of the most frequently encountered or most brightly coloured species, together with interesting facts on basic taxonomy, modes of life, and habitats. This is the first guide of this area to provide photographs of living butterflies, on vegetation, as they would appear in nature, as an aid to identification. The colours of the butterflies are, therefore, just as you would see them, rather than from a plate or photograph of a museum specimen.

Traditionally the northern limit of the West Indies is considered the Bahamian archipelago and three of the Greater Antilles, Cuba, Hispaniola, and Puerto Rico. The arcs of the Lesser Antilles (the Leeward Islands from Anguilla to Guadeloupe and the Windward Islands from Dominica to Trinidad) complete the span.

Nearly all guides to Caribbean butterflies do not include the rich fauna of Trinidad and its smaller neighbour, Tobago, which together have over 650 butterfly species. Although, biologically, these two islands represent an almost unmodified extension of the adjacent continent of South America, politically they are very much a part of the Caribbean and they receive many visitors eager to learn the natural history of the area. For this reason, I include some Trinidadian butterflies.

To this mix I have added the butterflies of Florida. This is because about 40 per cent of the 120 butterfly species of central and south Florida are common to the West Indies, probably because their host plants are also in the West Indies. For example, many of the tree species in South Florida are also widespread in the West Indies including Lignumvitae (*Guaiacum officinale*), mahogany (*Swietenia mahogani*), strangler fig (*Ficus aurea*), gumbo limbo (*Bursera simaruba*), torchwood (*Amyris elemifera*), and wild tamarind (*Lysiloma latisiliqua*).

The West Indies: The area and its butterflies

Surprisingly, although the West Indies lie wholly within the tropics, they muster fewer species of butterflies than continental Europe: 300 species against about 390. Central America, on an equivalent latitude to the West Indies, has seven times as many butterflies. Part of the explanation is that the land area of the West Indies amounts to only some 90 000 square miles (240 000 sq. km), similar to that of the United Kingdom. Larger areas generally have greater varieties of habitats and support more species than small areas. Larger areas also support more endemic species which have evolved there. The relatively small size of many Caribbean islands reduces the number of butterflies they support.

Another part of the explanation of the relatively low diversity of butterflies is that the Caribbean islands are volcanic in origin. The Caribbean fauna consists of samples of butterflies derived from the mainland, which, over an immense period of time, have succeeded in establishing themselves on the islands. Many insects fly and can be transported by high winds. The principal route of immigration was from Central America via Cuba, then in stepping-stone fashion into Hispaniola, Puerto Rico, and finally into the Lesser Antilles. Some species arrived from South America via Trinidad and Tobago, and then migrated into the eastern Caribbean. Especially in Florida, some species have emigrated from the northern United States.

Because most species emigrated from Central America via Cuba, and because larger islands support more species, the Greater Antilles support many more types of butterflies than do other Caribbean isles. Hispaniola, for example, has 151 species of butterfly, of which 41 are endemic. In large measure this is a result of the luxuriant radiation of *Calisto*, a genus of Satyrids that proliferates there. Smaller islands have fewer endemics. Dominica, in the Leeward Islands, has only two endemic species.

What is a butterfly?

Butterflies and moths belong to the class Insecta (the insects). They are members of the family Lepidoptera. Lepidoptera are flying land insects with complete metamorphosis, in which the development of the insect passes through the stages of egg, larva (or caterpillar), pupa (or chrysalis), and imago (or adult) with four wings covered with coloured scales. The adult usually feeds on nectar from flowers, with the aid of a suctorial proboscis. The caterpillar, which is equipped with biting mouth parts, usually feeds on growing plants.

There are several differences between butterflies and moths: moths fly mainly at night, butterflies in the day; while at rest, a butterfly folds its wings back over its back, a moth folds it wings horizontally so that it looks like a grounded aeroplane. Also, most moths spin a silken cocoon around the chrysalis; butterflies do not.

The body of a butterfly is divided into three basic sections: the head, the thorax, and the abdomen. Like other insects, the body is supported by a hard outer casing. Muscles and internal organs are attached to the inner surface of this outer casing, or exoskeleton.

The jaws (mandibles) in most adult butterflies have been highly reduced so that they have lost their original biting function. Other accessory mouth parts (maxillae) have been adapted for sucking purposes, being transformed into two elongated half tubes that are held together with hooks to form a tubular, hollow, sucking-and-licking organ, the proboscis. When not in use, the proboscis is coiled in a spiral. The proboscis may be as long as the adult's body.

The antennae project from the top of the head, and can be moved in all directions. They are usually threadlike, but in some families are thicker at the tip, or clubbed; others are bristle-shaped or comblike, or are intricately feathered. Delicate nerve cells that function as highly sensitive organs of smell are located here. The antennae are also used as sensory feelers.

The eyes of a butterfly are situated at the sides of the head. They are compound, or faceted, and built up from separate optical units, called ommatidia, of which a butterfly can have between 12 000 and 17 000. Butterflies seem to have excellent vision, particularly at shorter distances.

The thorax of the butterfly is distinctly separate from the head and the abdomen. It is attached to the head by a delicate, short, membranous neck, which gives the insect a partial ability to turn the head to the side. The thorax of a butterfly is composed of relatively strong segments that form a hard box, filled largely with muscles. It consists of three basic parts: prothorax, mesothorax, and metathorax. The forelegs are attached to the prothorax, and the middle legs and the pair of forewings are attached to the mesothorax. The metathorax bears the third pair of legs and the hindwings. The second and third thorax segments, which bear the wings, are firmly joined, thus giving strong support to both pairs of wings. On the sides of the thorax, two pairs of breathing holes (spiracles) are situated. In many species, butterfly feet, particularly the front pair, are sensitive organs of smell, through which the butterfly takes in the scent of nectar, flowers, or its sexual mate.

The soft limp wings of a butterfly newly emerged from its chrysalis are gradually expanded by pumping blood and air into the veins. The wings enlarge to their normal size and shape and, as soon as they harden, the butterfly is capable of flight. The distribution of wing veins is constant within individual species and is therefore useful for classification purposes. Butterflies beat both wings on either side simultaneously. The shape of the wings is roughly triangular, but in many families it varies considerably. The wings of the most powerful fliers, for instance, are narrowed; others are rounded. Sometimes the wing margins are variously cut out; at other times the wings form spurs or tassels, or are cleaved into feathery tails.

Butterflies are renowned for the exquisite colour of their wings. The scales give the colour to the wings. In the case of the matt-coloured species, the scales contain a variety of colour pigments which produce vivid or dull shades. These 'paints' are complex chemical compounds.

FIGURE 1   Mating malachites

Unfortunately, when these butterflies are handled the scales often rub off as a coloured powder. Other butterflies owe their beauty to physical factors, particularly to the interference of light in the fine layers of the wings. Among the many scales are sex scales called androconia, that lie in thick velvety patches. These produce scent hormones, or pheromones, which generate odours that attract mates and aid in sex recognition.

The abdomen of the butterfly is attached to the thorax and usually consists of ten well-defined segments. It is roughly oval in shape, rather soft and has no limbs or appendages. The last segments of the abdomen are usually joined and modified into sexual organs. In the soft membranous lateral parts of the segments there are six to seven pairs of respiratory openings called spiracles. The abdomen contains digestive organs, the heart and associated muscles, the excretory and sexual organs, and a complex muscular system. Species that do not feed in the adult state have the greater part of the digestive system filled with air, which serves to reduce the weight of the insect.

In females the genitalia are slitlike; males have a pair of grasping organs known as claspers. The internal sexual organs of a female take up the largest part of the abdomen. The main components are the pair of ovaries, each consisting of several egg-filled tubes. After fertilization the eggs become coated with a sticky secretion that enables them to stick to a selected leaf surface.

The male sexual organs consist of two testes that are fused together and produce the sperm. The sperm ducts carry the sperm toward the copulatory organ, the penis.

During mating, males and females may remain together for many hours. See Figure 1 – mating malachites.

The life cycle of butterflies

There are four stages in the life cycle of a butterfly: egg, caterpillar, pupa, and adult.

Egg

After fertilization, the female butterfly selects suitable places in nature for depositing her eggs (ova). The size of the egg fluctuates from 0.5 mm in diameter to a little over 3 mm, but is relatively constant in each species. The shape of the eggs varies between species and can be globular, disclike, conical, barrel-shaped, melon-shaped, bottle-shaped, or angular. The females of some species lay their eggs singly, other species cover the chosen area with neat rows. The eggs of the Giant Swallowtail may be laid together, *en masse* (see Figure 2).

The shells of all Lepidopteran eggs are chitinous. That is, they are covered with the tough, horny substance known as chitin, that forms the outside covering of insects; some are relatively soft, but more often are fairly hard, resilient, and firm, able to withstand changes of weather and surroundings, particularly during hibernation. The time it takes for a caterpillar to hatch can vary depending on temperature, humidity, and the actual species. Caterpillars of some tropical species hatch on the third day after the egg has been laid; whereas in some of the cooler zones of the world many species pass the winter at the egg stage, and it may take several months before the caterpillars hatch.

Caterpillar (Larva)

The second stage in the development of a butterfly is the larva, or caterpillar. Its body is elongate and is generally covered with a soft, flexible skin. The body surface may be covered in hair, sometimes sparsely, at other times thickly.

The caterpillar's body consists of as many as 13 or 14 segments. Three segments next to the head form the thorax. Each of these segments bears a pair of strongly hardened, jointed legs ending in a claw. The remaining segments are those of the abdomen. Some of these segments have a pair of false legs, or prolegs. The last of these are called claspers. The prolegs end in a contractile pad surrounded by a ring of hooks that are connected with muscles. The prolegs strengthen the grip of caterpillars on their host plants.

On the sides of the caterpillar's body are breathing holes (spiracles). In the forepart of the caterpillar's body are large salivary glands. The orifice of the salivary duct lies between the mandibles at the base of the head. The salivary glands produce a liquid that, when in contact with

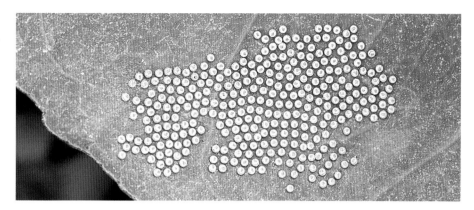

FIGURE 2   Eggs of the Giant Swallowtail

FIGURE 3   Regal Moth caterpillar larva – early stage

FIGURE 4   Regal Moth caterpillar larva – late stage, showing change in coloration from early larva

7

air, instantly solidifies into a strong, resilient thread called silk. The silk thread is essential to the caterpillar. Before each moulting it spins a pad of silk to which it attaches itself. Some use the silk to draw together the leaves in which they hide; others descend from trees with the aid of the silk thread, pupate in a silk cocoon, or use the thread to attach the loose pupa to a firm object.

The larva, during its development, may shed its skin four to five times as its appearance and coloration often change with the moulting (see Figures 3 and 4). The new, looser body covering allows for growth, so that eventually the caterpillar reaches its final size. At the end of the larval stage the caterpillar stops feeding, empties its gut, attaches itself to a pad, or crawls into the ground. After a certain period of rest, it sheds its skin and changes into a pupa.

Pupa (Chrysalis)

The sheath in which the future butterfly is formed is called a chrysalis or pupa. It is generally a fairly firm, hard, oval-shaped formation that narrows to a tip in the hind part. In the pupal stage the butterfly does not feed, does not moult and usually does not move. The transformation into the adult butterfly within the chrysalis occurs gradually, with the dissolution of some of the larval organic tissues, and the growth of new organs from tiny groups of cells known as imaginal buds.

It is possible to see most of the important parts of the future butterfly on the surface of the pupa. The hard covering of the eyes and legs on the front part of the pupa are often clearly mapped out. The future proboscis runs between the stretched-out legs, along the underside of the pupa, toward the hind end. The wings are also already marked out and some wing veins may be visible.

The shapes and coloration of pupae are as varied as the butterflies themselves. Contrast the photographs of the pupae of the Pipevine Swallowtail and the mirror-like *Tithorea harmonia*, known in Trinidad as the Tiger (Figures 5 and 6). The pupae of butterflies are particularly known for their characteristic hollows and grooves, ledges, hooks and horns, and sometimes beautiful colours. The emerging butterfly makes a hole in the thin cover above its head and inhales air through the mouth, filling its digestive system. A little later, as the skin of the pupa splits, the adult crawls out carefully, withdrawing its legs, antennae, and proboscis. When the legs are free, the butterfly stands up and extracts its wings out of the shell. Sometimes the adult will fail to extricate itself from the chrysalis, or one of the wings fails to expand properly (see Figure 7). These malformed individuals are destined to soon fall prey to some natural enemy.

FIGURE 5   Pipevine Swallowtail pupa

FIGURE 6   *Tithorea harmonia* pupa

FIGURE 7   Silver Spot butterfly – deformed after emergence from chrysalis

9

What do adult butterflies feed on?

Most adult butterflies feed on nectar from flowers. However, nectar and sweet juices from rotting fruit are not the only diet of the Lepidoptera. Some species are more inclined to suck the sap from wounded trees, while others, such as fritillaries, show a preference for dung, carrion, and rotting plant material. Butterflies also need water. They are fond of sucking it from damp, sandy, or muddy places. (See Figure 8). Some butterflies, such as *Vanessa*, will even alight on people to lick their sweat, which is rich in sodium. These butterflies appear tame and can often be caught easily and examined.

It may be possible to strengthen a weakened butterfly with nourishment. Place a few drops of diluted honey on the edge of a small dish, hold the butterfly lightly by its closed wings, uncurl the proboscis with a needle, then dip the proboscis tip into the liquid. The butterfly may start to suck. Sometimes it may start to feed independently. It can be induced to take clean water in the same manner.

Butterfly collectors are known to paste bait on stones and barks of trees in suitable natural localities where they capture the feeding butterflies. Generally these baits are formed from boiled mixtures of honey, syrup, and fruit, such as apples. The caterpillars of butterflies have a much more narrow menu; most species depend on a particular plant species and feed mostly on its leaves.

FIGURE 8   Sulfur butterfly drinking from sand

Protection and vulnerability

Batesian mimicry

Many butterflies are protected against birds and other insect-eating vertebrates by being permeated with toxic and distasteful chemicals. Other species that do not have these characteristics may resemble the latter in size, colour, and design. The resemblance to an obnoxious species is often sufficient to put off a predator. Such an occurrence is called Batesian mimicry.

Batesian mimicry owes its name to the British naturalist, Henry Walter Bates, who studied the phenomenon of mimicry in the South American jungles. He noted that some Heliconiidae, inedible and thus immune from attacks by animals, are copied by edible Pieridae. In other words, the latter so closely resemble the Heliconiidae that they are almost indistinguishable and are, therefore, also shunned by birds and other enemies. Pierids, of course, do not make any conscious effort to mimic the model, that is, to assume the form and colour of the Heliconiidae. Bates' good friend Alfred Russell Wallace (who, together with Darwin, proposed the theory of natural selection) put it this way:

'The number of species of insects is so great, and there is such diversity of form and proportion in every group, that the chances of an accidental approximation in size, form, and colour of one insect to another of a different group is considerable; and it is these chance approximations that furnish the basis of mimicry, to be continually advanced and perfected by the survival of those varieties only which tend in the right direction.'

Mullerian mimicry

Mullerian mimicry owes its name to a German naturalist living in Brazil, Fritz Muller. According to Mullerian mimicry, common coloration helps to educate enemies. Young insectivorous predators, not yet knowing what they may or may not eat, are apt to attack inedible insects, but will, after several encounters, stop making the same error. Since several species sport the same warning coloration, the relative loss that each of them suffers through such misguided assaults will be bearable for each. Mullerian mimicry thus turns out to be of advantage to all concerned.

E. B. Poulton cleverly characterised the difference between Batesian and Mullerian mimicry thus:

'A Batesian mimic may be compared to an unscrupulous tradesman who copies the advertisement of a successful firm; Muellerian mimicry, to a combination between firms to adopt a common advertisement and share the expenses.'

Camouflage

Other butterflies gain protection by not being visible. These often show no movement at all and do not attract a predator's attention. In such instances protective colouring, or camouflage, makes the butterfly merge with the bark of trees, foliage, or stones to such a degree that it is almost invisible. In these cases the insect's colours mimic the surroundings. Even the most exquisitely coloured butterflies may have their undersides in dull colours, so when at rest with their wings closed, they become inconspicuous. The design on the underside of the wings can imitate various types of tree leaves; green, dead or decaying ones.

On the other hand, some species of butterfly have their undersides most strikingly coloured, which can have a frightening effect on an enemy. For example, individuals of the genus *Caligo* which fly at dusk, have 'owl eyes' on the undersides. Their frightening appearance is coupled with the movement of wings. These butterflies love the twilight, when owls also set out hunting. Small insect-eating vertebrates, particularly the small climbing mammals and bats, also hunt at this time. All these creatures have a great respect for owls, and thereby may be warned off by seeing the picture of the owl's eyes on the underside of the butterfly.

The caterpillars of many Lepidoptera are also often well protected by camouflage. Some swallowtail caterpillars resemble bird chalk, or droppings, in their early stages. Other caterpillars are highly visible and probably chemically protected. The Monarch caterpillar has bright black and yellow stripes and is protected by cardiac glycosides gleaned from its host plant. The potently scented forklike protuberance (osmeteria) behind the head of the swallowtail caterpillar is also considered to be protective.

Vulnerability

Although butterflies are relatively well protected against larger enemies, they are prone to attack by predatory insects. In particular, parasitic wasps and flies can cause immense losses from time to time. These species lay their eggs in or on the body of the caterpillar or inside the egg or pupa. The developing wasp or fly then eats the caterpillar or pupa from the inside out, totally devouring its body contents.

In such cases mimicry or camouflage is useless as the parasitoids search by smell. In addition, bacteria, fungi, and viruses often invade caterpillars, causing disease. Parasitic wasps or viruses can be a particularly severe problem at commercial butterfly gardens where a great deal of stock may be destroyed by them.

West Indian and Floridian habitats

It is valuable to have an understanding of a butterfly's habitat requirements. The place where it lives is determined primarily by where its larval food plant occurs. Information on habitat types is a valuable signpost on what species to expect on what islands. Probably the richest habitat type for Caribbean butterflies is tropical forests. This is because rainforests have the most plant species. Because each butterfly is specific to one type of host plant, more plant species means more butterfly species.

Forests

Rain forests do not consist only of trees: a profusion of smaller plants grows on the ground below, as well as on tree trunks and branches. Many of the latter, such as the lianas or vines, are mechanically dependent on the trees for support. Epiphytes are another group of plants found growing on trees. These include a variety of algae, mosses, liverworts, lichens, and ferns, together with flowering plants such as orchids and bromeliads (see Figure 9), which attach to trunks or branches close to the ground or high up in the canopy. This profusion of different types of plants provides many different sorts of larval host plants for butterflies.

Small tracts of rain forest exist on almost every Caribbean island, modified to varying extents by plantation and forestry practices. They are the remnants of more extensive forest vegetation that covered much of the islands before the advent of humans. However, even in its natural state, the forest is not uniform in its development nor in the variety of plant species, because different types of forest are produced in different environments, depending on local variations in climate, topography, and soil type. Tropical rain forests exist in the Lesser Antilles on gently sloping, sheltered mountainsides up to 1000 m in areas which have rainfall in excess of 200 cm a year and constantly high humidity. In the Greater Antilles, the lower slopes of the Luquillo Mountains in north-east Puerto Rico (see Figure 10) and parts of central Hispaniola support rain forest, but relatively little survives in Cuba or Jamaica.

Rain forests on high Antillean mountains may be different from

FIGURE 9 Epiphytes growing on trees, typical of tropical forests

FIGURE 10 The Luquillo Mountains, Puerto Rico

those at lower elevations. This is because exposure to drying winds increases, particularly on exposed ridges, and at about 1000 m temperatures fall to levels similar to those of some temperate regions of the world. Montane rain forest, in which tree growth is poorer than in the lowland forests, is developed at this altitude. Summits have almost continuous precipitation because clouds cover them for most of the year, despite strong winds. Here an elfin woodland is developed, comprising a few species of short, profusely branching trees covered thickly with mosses, lichens, and other epiphytes, shading an understory of ferns and small palms. Such mountain forest types can be found in the inner arc of the Lesser Antilles, where high mountains with fertile slopes occur, such as on Mt Diablotin, Dominica; Grand Etang, Grenada; La Soufrière and Sans Toucher on Basseterre, Guadeloupe; and on St Lucia. In Puerto Rico, stands of the cabbage palm, *Euterpe*, occur between 650–800 m just below the zone of montane rain forest. In Jamaica, the montane rain forest is found much higher on the Blue Mountains. Common butterflies in tropical forests include shade-loving species such as the Mosaic, the Malachite, and various heliconiids.

15

Florida has no true tropical rain forests. There are no mountains for them to grow on and the soil is so sandy that the rain drains through quickly. However, there is some deciduous forest. Mixed broadleaf forests or 'hammocks' are common in moist conditions or near rivers or lakes (see Figure 11). Such forests exhibit the highest diversity of any communities in Florida. Vines and epiphytes, such as resurrection fern, are also common. The Fakahatchee strand, east of Naples, represents an extensive south Florida example of hardwood swamp. The tropical influence in this habitat is evident in trees festooned with epiphytic bromeliads, ferns, and a myriad of Caribbean and South American orchids. The Caribbean influence is pronounced with such trees as wild tamarind, poisonwood, pigeon plum, mahogany, gumbo limbo, and strangler fig, creating an almost West Indian landscape. In Florida, butterflies of the hammocks include the Ruddy Daggerwing, the Florida Purplewing, the Zebra Swallowtail, Schaus' Swallowtail, and the Giant Swallowtail.

FIGURE 11   Mixed broadleaf forest or 'hammocks' in Florida

FIGURE 12   Savannah at Aripo, Trinidad

Savannahs

Forests reach their best development on soils that are rich and well drained. In poorer or less well-drained soils, a savannah may be produced. The savannah flora consist largely of grasses, sedges, and other small herbaceous species, with scattered clumps of trees. Little natural savannah occurs in Jamaica or Puerto Rico, but savannah is found in Hispaniola and Cuba, although it is not as widespread as in former years. In Haiti and the Dominican Republic, savannahs are found particularly above 500 m. Where the soils are deeper, scattered groups of pines and other trees become established, but are stunted and low-crowned. More than one-third of the plains and gently undulating foothills of Cuba were originally under savannah vegetation.

Very little natural savannah vegetation exists today in the Lesser Antilles, although it was probably never very extensive even in earlier times. Areas of grassland do occur, as on the Grand Savannah in Dominica and some parts of Antigua, but these are almost certainly developed from degraded forest, largely as a result of fire. Natural savannah also occurs in south-east Barbuda, on patches of poorly drained soils scattered between clumps of trees, cacti, and small palms.

17

FIGURE 13   The sundew or *Drosera*, which grows on nitrogen deficient soil

FIGURE 14   Pine flatwoods in Florida

18

At Aripo, in north central Trinidad (see Figure 12), an iron-pan subsoil layer prevents water drainage, and a flora of sedges and grasses exists, producing an open savannah. The soils at many savannahs, including those in Florida, are poor in humus and nutrients, so some plant species rely on other food sources for their survival. Insect-eating plants, such as pitcher plants or venus fly traps, are present, utilising nitrogen-containing substances obtained from the digestion of their prey. One of the commonest of these is *Drosera*, the sundew, which bears sticky tentacles on its reddish leaves (see Figure 13). Any insect or small invertebrate that touches the tentacles is held fast and its struggles stimulate the other tentacles to bend over and secure it. The plant then secretes digestive juices from the leaves onto the prey, afterward absorbing the liquid products of digestion.

One of the most savannah-like habitats in Florida is pine flatwoods (see Figure 14). Flatwoods are characterised by sandy soils underlain by an impermeable layer (hardpan) that maintains a perched water-table during the rainy season. Flatwoods are usually dominated by longleaf pine and slash pine, but interspersed may be grassland. Summer fires set by frequent lightning storms keep pine flatwoods appearing rather open and parklike. Lightning strikes and fires are much more frequent in Florida than in most areas of the Caribbean. Scattered understory species include saw palmetto, gallberry, wax myrtle, and runner oak. Good examples of flatwoods can be found in the Apalachicola and Osceola National Forests. Cypress swamps may be interspersed within the flatwoods landscape. Butterflies of savannahs include various lycaenids and pierids, and especially in the tropics, the ringlets.

Arid areas

Other areas of Caribbean islands lie under a rain shadow of high mountains and may be considerably drier than windward shores or upland areas. Eastern and central Jamaica receive about 450 cm of rain annually compared with less than 100 cm in the Kingston area and the offshore cays. Whereas Dominica has over 500 cm per year on its high mountains, the much flatter island of Antigua is considerably drier with an annual average rainfall of only 115 cm. Barbados has a similar mean annual rainfall to that of Trinidad, but the lower relief, porous soil, and rocks and constant winds give this island a much drier environment.

Under these conditions a deciduous scrub formation is produced with low trees 3–4 m high. Cacti and agaves occupy many parts of the Lesser Antillean Islands, as at Hackleton's Cliff, Barbados, and Praslin, St Lucia. Such habitats are found also in south and west Puerto Rico on limestone hillsides, on the Portland Ridge in Jamaica, and on the islands off the west coast of Trinidad, where rainfall is less than 100 cm

FIGURE 15   Floridian coastal community

annually. In these areas the dry forest contains the locust tree, *Hymenaea* and acacias interspersed with *Cereus* and *Opuntia* cacti, and agaves. Butterflies of drier coastal areas include buckeyes, *Junonia* spp., pierids such as *Eurema* spp., *Phoebis sennae* and *Ascia monuste*, the Gulf Fritillary and the Painted Lady, *Vanessa cardui*.

In Florida, dry scrub occurs abundantly in the Ocala National Forest of central Florida and along the central Lake Wales Ridge. White, nutrient-poor, sandy soils, which drain quickly, support scrub oaks, scrub hickory, saw palmetto and rosemary. In sand pine communities, fires are infrequent and burn hot, devastating standing vegetation that ultimately returns to a sand pine dominated community. In turkey oak communities, fires occur more frequently and the dominant trees are longleaf pine and turkey oak, with wiregrass in the understory. Similar scrub may grow in coastal areas (see Figure 15).

FIGURE 16   Pickerelweed pond, typical of freshwater habitat

Freshwater marshes

Freshwater marshes require the regular influence of changing water levels to maintain their unique structure. A good example of this habitat in Florida is pickerelweed ponds (see Figure 16). Here a permanent drop in water-level allows the invasion of woody species. Freshwater marshes are often threatened by drainage and agricultural expansion. In the Caribbean, small areas of swamp are found in most of the Antillean islands, and some extensive swamps have developed at Zapata in Cuba, the Boqueron Valley in Puerto Rico, around the Black River in Jamaica, and on the east coast of Trinidad at Nariva. Here floating water plants like *Pistia* and *Salvinia* grow. Typical butterfly species found here include the Palamedes Swallowtail and the White Peacock.

21

How to study butterflies

In the past, most butterfly enthusiasts maintained collections of butterflies as part of their hobby. Today, however, many readers prefer to know butterflies by learning how they live, and by watching and photographing them. Butterfly behaviour in some species is still little known; here is an area where observers can make genuine contributions. The general rules for butterfly watching and photography are essentially the same: haunt sunny, flowery places, be observant, and move very quietly and slowly. Patience is definitely required. Many of the photographs in this book were taken with a 105 mm macro lens, sometimes with a ring flash.

Another way to enjoy butterflies is to rear them from eggs or caterpillars. The caterpillar or egg may be easily brought in from its natural habitat. Large clear or clear-topped plastic boxes can be used to keep leaves and small branches of the caterpillar's host plant. Regular cleaning minimises the risk of disease. When the caterpillar has finished feeding it will often pupate on the twigs. Twigs also provide a good foothold for the emerging adult. After the new butterfly emerges and dries its wings, it can be kept or returned to its natural environment.

By cultivating nectar flowers and host plants it is also possible to attract many butterflies to your garden. A number of native butterfly species may become established quickly, and others will visit flowers for nectar, if not to breed. A good start can be made by growing some passion flowers and *Citrus* in a chosen area, and then planting *Buddleia* and *Lantana* in between seasons. This arrangement will attract heliconiids and swallowtails to the garden to breed and feed. Knowing a chosen caterpillar's host plant, as well as the adult butterfly's preferred nectar source, will help you determine which plants to cultivate. You will find larval host plant information in the individual species accounts.

In Florida and the Caribbean there are many flowers that are attractive to butterflies as nectar sources. A list of these is provided in Table 1. Some of the best include *Buddleia* spp., the so-called butterfly bush, and butterfly weed or milkweed, *Asclepias tuberosea*. There are more than 50 species of *Buddleia*, and about eight or nine have been cultivated in Florida at one time or another. The most widely grown is *Buddleia davidii*, which will grow up to 6 m tall. The varieties of this plant have purple, lilac, pink, red, or white flowers. Plants can be

Table 1 **Flowers attractive to butterflies (after Stiling, 1989)**

Common name	Scientific name
Trees	
Bay cedar	*Suriana maritima*
Bottlebrush tree	*Melaleuca quinquenervia*
Chinaberry	*Melia azedarach*
Citrus	*Citrus* spp.
Poisonwood	*Metopium toxiferum*
Shrubs	
Butterfly bush	*Buddleia lindleyana*
Fetterbush	*Lyonia lucida*
New Jersey tea	*Ceanothus americanus*
Red buckeye	*Aesculus pavia*
Sassafras	*Sassafras albidum*
Christmas bush	*Eupatorium ordoratum*
Pentas	*Pentas* spp.
Herbs, vines, and ground cover	
Blazing star	*Liatris tenuifolia*
Butterfly weed or milkweed	*Asclepias tuberosa*
Clover	*Trifolium* spp.
Dotted horsemint	*Monarda punctata*
Goldenrod	*Solidago fistulosa*
Honeysuckle	*Lonicera sempervirens*
Ironweed	*Vernonia angustifolia*
Lantana	*Lantana camara*
Pickerelweed	*Pontedaria cordata*
Spanish needles	*Bidens pilosa*
Thistles	*Cirsium* spp.
Tickseed	*Coreopsis gladiata*
Verbena	*Verbena brasiliensis*
Stachytarpheta	*Stachytarpheta* spp.

propagated by seed or cuttings, but grow spindly unless pruned.

The butterfly weed, *Asclepias tuberosa*, has yellow to orange-red flowers, which are attractive to many species of butterfly. The related *A. curassavica* is more common in the tropics (see Figure 17) but has been introduced into Florida. Plants are best purchased from commercial nurseries. Of the other species listed, it is worth remembering that *Lantana* is poisonous, especially the berries, which are toxic to children. This feature is unfortunate because it is one of the best natural

FIGURE 17   Butterfly weed (*Asclepias curassavica*)

butterfly plants in Florida and the Caribbean. *Bidens* has sticky seeds and thistles, is prickly, and may be an inconvenience.

It will probably be difficult to establish all of the flowers listed in Table 1 in your garden, but a good rule of thumb is that sweet, pungent, and acid-smelling flowers attract butterflies. Particularly attractive colours include orange, yellow, pink, purple, and red. White flowers and those emitting fragrance at night tend to attract moths. Plants with deep-throated, drooping, or enclosed flowers are not well suited for nectar gathering. Wild flowers are great attractors; many hybridised flowers are not.

Providing host plants for butterfly caterpillars is a little more difficult and a little more unpopular because the plants involved are sometimes considered weeds. Furthermore, only nine per cent of butterfly species are polyphagous, that is, are able to utilise a wide variety of plants. Most are monophagous or highly specific, using only one kind of plant or a very narrow range. Two sweepstake plants that might be considered are *Cassia* spp. and *Passiflora* spp. Passion flowers, (*Passiflora*), can provide a nectar drink and are host to three attractive butterflies, *Agraulis vanillae*, the Gulf Fritillary; *Heliconius charitonius*, the Zebra Longwing; and *Dryas iulia*, the Flambeau. *Cassia* spp. are host plants for several species of sulfur butterflies.

24

Butterfly collecting

Collecting butterflies is still a very important part of Lepidoptera research; nevertheless, those who collect for study should follow laws concerning endangered species and rules of common sense and consideration for the resource. Without these guidelines, we may lose some of our most treasured butterfly species.

Butterfly nets are available from dealers in natural history apparatus in a variety of shapes, sizes, and prices, and many of them are made to fold up or take apart for easy packing. Most are made of dark-coloured, tough, but soft netting and taper a little at the bottom to restrict the butterfly's movements.

Killing jars can be made by pouring about 12.5 cm of plaster of Paris into a wide-mouthed jar. After the plaster has dried thoroughly, saturate it with ethyl acetate. When the jar loses its killing power, the chemical treatment can be repeated. Alternatively, some collectors simply pinch the butterfly's thorax between thumb and forefinger to kill it. This alleviates the problem of carrying heavy apparatus around. Strips of rough paper placed in the jar will reduce damage to specimens. For temporary use, you can use a jar containing a few paper strips moistened with ethyl acetate. Specimens too stiff for mounting can be relaxed by placing them in an airtight jar containing a layer of moist sand or moist paper. This relieves the stiffness of rigor mortis.

Spreading boards are made of soft wood with a centre channel into which the body of the butterfly fits. The specimen is pinned through its thorax and its wings are spread at right angles to the body. To reduce the loss of scales, forceps or an insect pin should be used. The wings are then held in place with paper strips. Allow several days for drying. Spreading boards can be purchased or made at home.

Insect pins are made of special rust-resistant steel and come in several sizes. Size 3 can be used for most butterflies. Purchase them from biological supply houses.

Labels should be placed on the pin of each specimen. They should give information on where, when, and by whom the specimen was taken. Labels should be neat and small. When much collecting is done at one place, labels printed in advance are useful. These can be ordered from a biological supply house. Sheets of typewritten labels can be reduced to make very small but readable labels.

The species label, containing the scientific name of the specimen, is larger and is usually pinned to the bottom of the collection box by the specimen pin. Each species in this guide is given two names, a common name and a scientific name. The former is usually a national, the latter an international, name. Most of the former have been in existence for a long time. Others are relatively new or are regional. In this book, Gerberg and Arnett (1989), Riley (1975), and Barcant (1970) have been used for common names. Although common names may vary from country to country, the scientific (Latin) names never differ, and herein lies their value.

Storage and display boxes are of several kinds. The Riker mount, a cotton-filled, flat cardboard box with a glass top, is used for display. For storage of pinned specimens the Schmitt box, with cork bottom, glazed paper lining, and tight-fitting lid, is ideal. Cigar boxes fitted with balsa or corrugated cardboard bottoms, or even clear plastic boxes at least 5 cm deep, are good. All boxes must be treated periodically with paradichlorobenzene or naphtha moth crystals, to keep out destructive pests. Schmitt boxes have a special slot in which to pour the crystals; for other boxes a little perforated paper envelope will work well.

Commercial butterfly gardens and state insect collections

For those interested in viewing beautiful butterflies from around the world, commercial butterfly gardens represent a wonderful opportunity to see the staggering variety of colours and patterns exhibited by all manner of butterflies. Unfortunately, there are no butterfly gardens in the Caribbean but on 25 March 1988, the first butterfly farm in Florida, indeed in North America, opened at Butterfly World in Fort Lauderdale, Broward County. Butterfly World is located at Tradewinds Park, 3600 W. Sample Road, Coconut Creek, Florida 33073.

'Wings of Wonder, the Butterfly Conservatory' is a spectacular 5500 sq. ft conservatory in a rain forest setting at Cypress Gardens, in Winter Haven, Florida. More than 50 butterfly species from around the world grace a variety of plants. Many Caribbean species abound including swallowtails, owl butterflies, morphos, sulfurs, and the Zebra Longwing. Many other Florida species can be seen on the wing in the famous gardens themselves (see Figure 18).

For those in north Florida, another convenient display of living butterflies is provided at Callaway Gardens, Pine Mountain, Georgia, near Columbus. Open since the autumn of 1988, it exhibits over 50 species of free-flying tropical butterflies inside a large glass-enclosed butterfly conservatory. In addition, 65 native species can be found in the beautiful and extensive gardens.

FIGURE 18   Cypress Gardens, Winter Haven, Florida

Butterfly conservation

Although butterflies are adapted to defend themselves against natural enemies by means of camouflage, noxious taste, or mimicry, they are less well adapted to deal with the effects of human actions.

The rapid expansion of human activities in modern times has reduced the amount of natural vegetation available for wildlife in general. Many butterfly host plants are rarer than they were in former times. Furthermore, the use of highly toxic pesticides has also led to considerable reduction in the numbers of some butterflies in recent years. Spraying for mosquitoes on north Key Largo, Florida, has done much to decrease the already small numbers of Schaus' Swallowtail living there. Few pesticides are so specific that they will kill mosquitoes but not butterflies.

Incredibly, adaptations by Lepidoptera to these new conditions are sometimes possible. One of the well camouflaged European moths is the Peppered moth (*Biston betularia*). This moth has black-and-white mottled markings on the wings which make it virtually impossible to see on lichen-covered tree trunks. Trees in industrial regions get coated in soot and are blackened. On such tree trunks the black-and-white mottled moth stands out and can easily be caught by predators, such as birds. A black mutation of *Biston betularia*, normally picked off on lichen-covered trees, has an advantage on sooty trees and is protected from predators. Therefore, in industrial regions, *Biston betularia* is well represented by the all black or carbonaria form, while in the country, away from industry, the ordinary peppered form exists. Now that industrial pollution is being reduced, the black form is becoming much less common.

In addition to the effects of habitat reduction and pollution, many authorities have noted how overzealous collecting of adults may destroy populations. The Large Copper butterfly, *Lycaena dispar*, discovered in marshy areas of Britain in 1790, had been collected to extinction by 1849. Butterfly collecting is not always based on scientific interest. The beauty of butterflies has tempted many into the manufacture of many kinds of souvenirs, made up either of the whole butterfly or only of the wing of the various vividly coloured species. This is particularly true in Asia and in South America where souvenirs adorned with hundreds of Morpho wings may be bought.

Of the three factors, over-collecting, pollution, and habitat destruction, the latter has the greatest effect on butterflies. The species which inhabit forested areas have managed to survive in fewer and fewer localities as these areas are cleared for agriculture. Modern arable practice has reduced the stands of some wild plants to small localised patches. Butterflies which require a patchily distributed plant species, either for food or to lay eggs on, have a hard job locating the few remaining plant stands.

Unfortunately, refuges where some remains of the original vegetation still remain are rare, particularly in heavily populated areas where the demand for space is so great. Barbados, an island largely cleared for sugar cane planting by the end of the seventeenth century, has a paltry 24 butterfly species. Seminatural vegetation occurs only at Turners Hall Wood and on parts of the coast. In Florida, growing cities and suburbs are making significant impacts on the state's wildlife. Nearly 8000 new residents move to Florida each week.

Do your best to preserve everything natural that can still be preserved, so that future generations can still experience the pleasure of seeing a butterfly in flight over a habitat filled with colourful flowers.

THE BUTTERFLIES

Milkweeds: Danaidae

The Danaidae, predominantly a tropical and subtropical family of butterflies, are referred to as Tigers, in reference to their striking orange coloration with severe black stripes. This pattern is an advertisement of noxious taste, the noxious element consisting of heart poisons (cardiac glycosides) accumulated from the food plants, milkweeds (Asclepiadaceae and Apocynaceae) by the larvae and stored by the adults. The use of the same basic warning coloration by many species of danaids is known as Mullerian mimicry and tends to deter predators from attacking any prey species displaying it. Although there are about 300 species of milkweed butterflies worldwide, only nine are found in Florida and the Caribbean. The pupa is stout, usually green and with gold spots.

The Monarch, Queen and Large Tiger are shown in Figures 19, 21 and 23. Other danaids in the Caribbean are *D. cleophile*, the Jamaican Monarch, which looks like a small monarch but has yellow spots on its wings instead of white ones; and *D. eresimus*, the Soldier, which is very similar to the Queen but has a series of very faint spots, only slightly paler than the ground colour, on the underside of the hindwings.

Monarch: *Danaus plexippus* FIGURE 19, 20

The Monarch is a large butterfly with an 8–10 cm wing-span, bright orange wings with black veins, and no black line across the hindwing. The similar Viceroy is smaller and has a black line across the hindwing.

The Monarch caterpillar is off-white with black and yellow transverse stripes and has a pair of black filaments extending from the front and rear ends (see Figure 20).

Normal larval hosts are poisonous milkweeds, which give the larva and the adult unpleasant taste. Predators such as birds learn to associate the bright coloration of the Monarch with this bad taste. Recently some alternative hosts have been found, such as dogbane (*Apocynum*), which contains no noxious chemicals.

The adult hatched from the larva feeding on these plants is not so poisonous; it therefore relies solely on its resemblance to noxious

FIGURE 19   Monarch (*Danaus plexippus*)

FIGURE 20   Monarch (*Danaus plexippus*) caterpillar

individuals (automimicry) for protection. The adult prefers to take nectar on flowers of the larval food plant but it will also visit ornamental plants such as *Zinnia*, *Bougainvillea* and *Bidens*.

Often seen on the wing in October, the Monarch undertakes long migratory journeys from northern areas in the United States to overwintering sites in Mexico, passing through Florida on the way. Monarchs cannot tolerate cold weather, although Florida may be sufficiently warm for some communal overwintering roosts to form.

The Monarch's strong flight ability probably enhances its almost circumtropical distribution; it can be found in North and South America, the Canary Islands, Indonesia, and Australia. Some Monarchs tagged in Ontario have been known to fly 1870 miles in 129 days on their journey southward to Mexico. Populations of Monarchs in the western United States may overwinter on the coast between Monterey and San Diego. For example, John Steinbeck refers to overwintering roosts at Pacific Grove in his novel, *Sweet Thursday*.

In the Spring, Monarchs make the return journey north so that larvae can utilise the abundant milkweeds that grow in the United States.

Caribbean Monarchs are resident and do not migrate. In fact, several Caribbean subspecies are recognized and each island could have its own distinctive form of the Monarch.

- Wingspan: 90–100 mm
- Range: Through the Americas and the West Indies; circumtropical
- Flight period: Year round, but rare between May and October in south Florida
- Larval host plants: Milkweeds (*Asclepias*), dogbane (*Apocynum*)
- Caterpillar: See Figure 20

Queen: *Danaus gilippus*   FIGURE 21, 22

Possibly more abundant in Florida than the Monarch is the closely related Queen butterfly. It is also large but has a deep brown base coloration (as opposed to orange), with black margins and fine black veins. It is rare in the Lesser Antilles. Throughout its range, this is a butterfly of open land, including meadows and marshes. The adult feeds on asclepiads, *Lanata*, *Croton* and other flowering weeds. Caterpillars feed on asclepiads, including white vine (*Sarcostemma clausa*) and oleander (*Nerium oleander*).

- Wingspan: 75–85 mm
- Range: The south-eastern United States, Caribbean, Central and South America to Argentina
- Flight period: Year round
- Larval host plants: Commonly on milkweeds
- Caterpillar: See Figure 22

Large Tiger: *Lycorea cleobaea*   FIGURE 23

The Large Tiger is a forest butterfly often found close to streams or tracks, though extending into plantations. It is common in the Greater Antilles but much rarer in the Lesser Antilles, probably as a stray from

FIGURE 21   Queen (*Danaus gilippus*)   FIGURE 22   Queen (*Danaus gilippus*) caterpillar

FIGURE 23   Large Tiger (*Lycorea cleobaea*)

South America or Trinidad. In Costa Rica, it appears to mimic tiger-striped members of the genus *Heliconius* and *Melinaea*, and in South America and Trinidad the Ithomid genus, *Mechnanitis*, looks very similar. However, these genera are absent in the Caribbean. The only candidate for a model in the Greater Antilles is *Eueides melphis*, a rare insect of tropical forests.

- Wingspan: 96–100 mm
- Range: Mexico to Argentina, Caribbean
- Flight period: Unknown
- Larval host plants: Commonly on milkweeds and, surprisingly, figs (*Ficus* sp.)
- Caterpillar: Pale green, with wide black stripes and paired black fleshly tubercles on the second segment

Clearwings: Ithomiidae

The Clearwing butterfly family is exclusively neotropical, other than a single genus from the Australian region, and contains nearly 400 species of which only a few occur in the Antilles. They are very long-winged butterflies. They prefer the deeper, darker parts of the forest and venture out only rarely.

Two genera are common in the West Indies, *Greta* on the Greater Antilles and *Ithomia* on Trinidad. Both appear very similar. The two species of *Greta* are *G. cubana* on Cuba, and *G. diaphana* on Jamaica and Hispaniola. *Ithomia* occurs in Trinidad and Tobago. They are among the few West Indian butterflies on these islands with transparent wings; the forewing is elongated and the flight is very weak and laboured.

The larvae feed exclusively on Solanaceae, perhaps sequestering toxins since the adults are involved as models in many mimetic associates.

The caterpillars that are known are slender, smooth, pale in colour, and inconspicuously striped.

Blue Transparent: *Ithomia pellucida*   FIGURE 24

The Blue Transparent is one of the well-known window-pane butterflies (Ithomiidae) of South America, which may also be found in Trinidad and Tobago. The wing veins stand out finely in black, but the wing membrane is otherwise totally transparent, like a window-pane. In the field, under the darkened shade of the forest, the effect is slightly bluish. Little is known of its life history.

- Wingspan: 45–55 mm
- Range: South America, Trinidad and Tobago
- Flight period: Year round
- Larval host plants: Unknown
- Caterpillar: Unknown

Ringlets: Satyridae

The Ringlets are generally brown butterflies which are easily recognized by the eye-spots, or ocelli, that are almost invariably found at least on the underside of the hindwings. The larvae, which are smooth and spindle-shaped, have forked tails and feed on monocotyledonous plants. The adult butterflies are denizens of open grasslands. There are about 1500 satyrid species; more than 50 occur Florida and the West Indies. Eight species from five different genera occur in Florida, whereas all 40 of the Caribbean species belong to the genus *Calisto*.

FIGURE 24   Blue Transparent (*Ithomia pellucida*)

FIGURE 25   Night (*Taygetis echo*)

Hispaniola is the stronghold of *Calisto*, with just a few species known from Cuba and one each on Jamaica and Puerto Rico. The species are generally weak fliers with an erratic, bouncing flight.

Night: *Taygetis echo*   FIGURE 25

The Night butterfly is a South American Ringlet that also occurs in Trinidad and Tobago, and is typical of the appearance of most satyrids – dull brown on the upper wings. Like other members of the family, individuals prefer shady, moist conditions and are attracted to decaying vegetation and rotting fruit. Flight is low to the ground and settling is frequent, either on the ground itself or on low vegetation with wings folded. Most *Taygetis* species fly in the later afternoon or at dusk.

- Wingspan: 68–75 mm
- Range: South America, Trinidad and Tobago
- Flight period: Unknown
- Larval host plants: Unknown
- Caterpillar: Unknown

Wood Nymph: *Cercyonis pegala* FIGURE 26

The Wood Nymph is common only in north Florida, becomes rare in central Florida, and is not found south of a line from Tampa on the west coast to Fort Pierce on the east coast. It can sometimes be found in open oak or pine woodlands. It exhibits the characteristic eye spots of satyrids on the underside of its hindwings.

- Wingspan: 50–70 mm
- Range: Eastern United States
- Flight period: June–July
- Larval host plants: Various grasses
- Caterpillar: Unknown

Lady Slipper: *Pierella hyalinus* FIGURE 27

The Lady Slipper is found most frequently in shady undergrowth of tropical forests. This butterfly flies close to the ground and often alights on rotting fruit and decaying vegetation. Not much is known about its life history.

- Wingspan: 60–80 mm
- Range: Northern South America and Trinidad
- Flight period: Year round
- Larval host plants: Unknown
- Caterpillar: Unknown

Fritillaries: Nymphalidae

The Nymphalidae are often referred to as the brush-footed butterflies because their unifying characteristic is a pair of vestigial forelegs, useless for walking but often dense with scales. There are few other common features of the family; wing shape, size, and colour come in a staggering array. These are medium to large butterflies with a fondness

FIGURE 26   Wood Nymph (*Cercyonis pegala*)

FIGURE 27   Lady Slipper (*Pierella hyalinus*)

for sunshine. They are often swift fliers and brightly coloured. Many are attracted to rotting fruit, carrion, and animal excreta. There are over 3000 nymphalid species worldwide, with 23 species breeding in Florida and about another 50 in the Caribbean.

The caterpillars are mostly spiny and feed on a wide variety of host plants, but most are dicotyledonous, in contrast with the food plants of

FIGURE 28   Silverking (*Archaeoprepona demophoon*)

FIGURE 29   Ruddy Daggerwing (*Marpesia petreus*)

FIGURE 30   Ruddy Daggerwing (*Marpesia petreus*)
caterpillar

FIGURE 31   Ruddy Daggerwing (*Marpesia petreus*)
pupa

38

satyrids. Chrysalids are usually thorny or angular in appearance and hang upside down from silken pads.

Silverking: *Archaeoprepona demophoon* (dead specimen)   FIGURE 28

The Silverking is a beautiful and charismatic denizen of forest clearances. It occurs on Cuba, Hispaniola, Puerto Rico and Trinidad. It is a widespread, but rare, butterfly and one of the fastest fliers in the Caribbean. Most specimens are very shy and can only be observed from a distance as they sit, with wings open, high up on tree trunks or leaves, with their heads pointed downwards. The Silverking is not common, perhaps due in part to high rates of parasitism.

- Wingspan: 100–110 mm
- Range: Mexico to Brazil
- Flight period: Unknown
- Larval host plants: *Mollinedia laurina* and possibly *Igna* species
- Caterpillar: Unknown

Ruddy Daggerwing: *Marpesia petreus*   FIGURE 29, 30, 31

At first glance the long tails of the Ruddy Daggerwing make it look like a swallowtail, although the coloration is similar of that to the Julia with which it often flies. In Florida, the Daggerwing can be found from Fort Lauderdale southward, most commonly in March, April, and May. Hardwood hammocks and thickets in the Everglades are a good place to search. Even within the city of Miami and in the Florida Keys it may be found in wooded enclaves. It may often fly high around *Ficus* tree host plants. In the Caribbean it is found on the Lesser Antilles and Puerto Rico, but is replaced on Hispaniola, Cuba, and Jamaica by the very similar *M. eleuchea*.

- Wingspan: 65–75 mm
- Range: Mexico to Brazil, Caribbean, south Texas, and south Florida
- Flight period: Year round, but common in March–May
- Larval host plants: Ficus plants; banyon tree (*Ficus citrifolia*); common fig (*F. carica*).
- Caterpillar: See Figures 30 and 31

Common Daggertail: *Marpesia chiron*   FIGURE 32

A skittish flier of forest clearings, the Common Daggertail is found only rarely in the Florida Keys and Jamaica, suggesting that individuals found there are strays from populations in Central America or

Cuba, its Caribbean stronghold. It is most commonly observed at fallen fruit or, as Figure 32 shows, on damp soil or sand.

- Wingspan: 50–60 mm
- Range: Cuba, rarely in Hispaniola, southern Texas to Brazil
- Flight period: July
- Larval host plants: *Ficus, Morus* and *Artocarpus* spp.
- Caterpillar: Yellow-orange with red transverse streaks and two black dorsal streaks, black spines on back

Mosaic: *Colobura dirce* FIGURE 33, 34

The Mosaic commonly can be found resting, head down, on trees in the heavy shade of forests or plantations. It has a wild flight of great speed between its rests on trunks. Common in Cuba, the Dominican Republic and Puerto Rico, it is rarely found at flowers and, if not resting on trees, can be seen on rotting fruit. The early instar larvae are gregarious and are protected by day in a tent made from a folded leaf. As Figure 33 shows, many specimens often lack the lower portion of the hindwing, marked on the underside with an eye, as a result of lizard or bird attacks.

- Wingspan: 66–72 mm
- Range: Mexico to Paraguay; Greater Antilles
- Flight period: Unknown
- Larval host plants: *Cercropia* trees
- Caterpillar: See Figure 34

Orion: *Historis odius* (dead specimen) FIGURE 35

The Orion is a large butterfly whose wings have velvety, dark brown uppersides and cryptically coloured undersides. At rest, with the wings folded, the Orion blends into the background, which is often tree bark. Ranging from Cuba throughout the Greater and Lesser Antilles, it is a strong flyer and is found on all the islands in a variety of habitats. It often flies high in the air. The related *H. acheronta* (or Cadmus) has five or six small white subapical spots, instead of one, like *H. odius*. The Cadmus occurs only in the Greater Antilles.

- Wingspan: 110–130 mm
- Range: Mexico to Argentina
- Flight period: Unknown
- Larval host plants: *Cercropia peltata* trees
- Caterpillar: The light green caterpillar is 75 mm long when fully grown, with light brown transverse markings. The head is red-brown with short, black, spiky horns

FIGURE 32   Common Daggertail (*Marpesia chiron*)

FIGURE 33   Mosaic (*Colobura dirce*)

FIGURE 34   Mosaic (*Colobura dirce*) caterpillar

FIGURE 35   Orion (*Historis odius*)

41

FIGURE 36   Cracker (*Hamadryas feronia*)

FIGURE 37   Queen Cracker (*Hamadryas arethusa*)

FIGURE 38   St Lucia Mestra (*Mestra cana*)

Cracker: *Hamadryas feronia* FIGURE 36

The Cracker is aptly named because, in common with all the Hamadryas species, it makes a click when flying. The noise is probably a distraction to predators because the butterflies can fly silently. Within the Caribbean, *H. feronia* occurs only in Trinidad, but the closely related *H. amphichloe* can be common in Haiti and the Dominican Republic, and is also known in Cuba and Jamaica. Both species rest on tree trunks, head downwards and with wings pressed against the surface, where their camouflage is very effective.

- Wingspan: 60–70 mm
- Range: Mexico to Paraguay
- Flight period: Year round
- Larval host plants: *Dalechampia*
- Caterpillar: Grey-green with a pale yellow line on the side, head with pair of long horns, body with many branched spines

Queen Cracker: *Hamadryas arethusa* FIGURE 37

A rare butterfly of elevated tropical forests, the Queen Cracker has beautiful velvet black coloration with iridescent blue spots sprinkled all over the wings. It is sometimes attracted to rotting fruit on the ground. Very little is known about its life history.

- Wingspan: 65–75 mm
- Range: Mexico to Brazil
- Flight period: Unknown
- Larval host plants: Unknown
- Caterpillar: Unknown

St Lucia Mestra: *Mestra cana* FIGURE 38

The St Lucia Mestra is a fragile-looking butterfly and a slow flier that settles on flowers bordering hill tracks and roadsides. In the Caribbean, it is found only from Trinidad through to Dominica. The related *M. dorcas*, which has orange coloration on the apical third of the forewing, is restricted to Jamaica.

- Wingspan: 40–50 mm
- Range: Northern South America and the Lesser Antilles
- Flight period: July–August
- Larval host plants: Possibly *Dalechampia scandens*
- Caterpillar: Unknown

Florida Purplewing: *Eunica tatila* FIGURE 39

The Florida Purplewing is instantly recognisable by the blue-purple iridescence on the upper surface of the wings. No other Florida butterfly has such a metallic sheen. For this colour to be visible, the butterfly must be in sunlight; otherwise it looks merely brown. The underside of the wings is bark-coloured, and purplewings are well camouflaged at rest on tree trunks. In Florida, purplewings are found only in the hardwood hammocks of Dade and Monroe Counties. They are more common in hammocks in the Florida Keys. The purplewing rarely visits flowers.

The Dingy Purplewing, *E. monima*, is found in similar habitats in south Florida but has much less purple coloration in the wings. Both species are also found in the Greater Antilles.

- Wingspan: 40–50 mm
- Range: South Florida, Central America and the Greater Antilles
- Flight period: Year round
- Larval host plants: Crab wood, *Gymnanthes lucida*
- Caterpillar: Unknown

Trinidad Admiral: *Adelpha cytherea* FIGURE 40

There are perhaps half a dozen species of *Adelpha* butterfly in the Caribbean and most share the characteristic of an orange patch at the apex of the forewing that is shown by few other Antillean butterflies. One species (*A. abyla*) occurs in Jamaica, another (*A. iphicla*) in Cuba, another (*A. lapitha*) in Hispaniola, and one (*A. gelania*) in Puerto Rico which is also present in the Dominican Republic. The species shown in Figure 40 comes from Trinidad and Tobago where another 10 species are known.

- Wingspan: 40–55 mm
- Range: Tropical South America
- Flight period: Unknown
- Larval host plants: Possibly *Gonzala* spp.
- Caterpillar: Dull brown and very spiny

Viceroy: *Basilarchia archippus* FIGURE 41, 42

A master of mimicry, the adult Viceroy was long thought to gain protection from its similarity to the unpleasant tasting Monarch, which most birds ignore. It was thus a Batesian mimic. The chrysalis mimics a bird dropping and the young hibernating caterpillar hides among leaves on host willows. It is abundant throughout Florida, but is

FIGURE 39   Florida Purplewing (*Eunica tatila*)

FIGURE 40   Trinidad Admiral (*Adelpha cytherea*)

FIGURE 41   Viceroy (*Basilarchia archippus*)

FIGURE 42   Viceroy (*Basilarchia archippus*) caterpillar

45

FIGURE 43   Red-spotted Purple (*Basilarchia astyanax*) – side and top views

FIGURE 44   Mimic (*Hypolimnas misippus*)

46

smaller than the Monarch, and has heavier black lines on the wings and an additional black line crossing the hindwing. In south Florida, Viceroys were argued to mimic the Queen butterfly more commonly than the Monarch. However, recent research has shown that, in Florida populations, the Viceroy was as unpalatable to birds as the Monarch itself, and more unpalatable than the Queen. Thus the mimicry was Mullerian.

- Wingspan: 65–75 mm
- Range: Canada, United States and Mexico
- Flight period: Year round
- Larval host plants: Willows (*Salix* spp.)
- Caterpillar: See Figure 42

Red-spotted Purple: *Basilarchia astyanax* FIGURE 43

At first glance, the Red-spotted Purple appears much like a swallowtail (especially a Pipevine or Dark Tiger Swallowtail) because it is large and has a dark colour with blue iridescence on the hindwings. It also has brick-red spots on the underwing. The species is a nymphalid and is thought to mimic the Pipevine Swallowtail to gain protection from the Pipevine's bright coloration, which warns of toxicity. A prerequisite for mimicry is, of course, that model and mimic be common over the same range and in the same habitat. Thus, like the Pipevine, the Red-spotted Purple is common throughout most of the eastern United States, ranging from open woodlands to forest edges. In Florida, it is common in only the northern part.

- Wingspan: 75–85 mm
- Range: United States and Mexico
- Flight period: March–October
- Larval host plants: Willows (*Salix* spp.), cherries (*Prunus* spp.), and hawthorns (*Crataegus*)
- Caterpillar: Virtually identical to the Viceroy caterpillar

Mimic: *Hypolimnas misippus* (dead specimen) FIGURE 44

Distributed worldwide in the tropics, the female Mimic resembles varieties of the Old World's *Danaus chrysippus* with coloration not dissimilar to *Danaus gilippus*, the Queen. Males, however, are velvety black, as shown in Figure 44, with iridescent purple encircling large white spots. Because of the resemblance to African danaids, many authors suggest that the Mimic was actually introduced into northern South America via the slave trade. From there it spread sporadically into Trinidad and the Lesser Antilles. On occasion it may become more common on certain islands following a hurricane from Africa, depending on which islands lie in the hurricane's path. It is a rare insect, at all times, in the Caribbean.

- Wingspan: 50–75 mm
- Range: Old World tropics and subtropics
- Flight period: Year round, sporadic
- Larval host plants: Not generally in the Caribbean
- Caterpillar: Not generally in Florida or Caribbean

Buckeye: *Junonia coenia* FIGURE 45, 46

The characteristic features of Buckeyes are their large eye spots with iridescent blue and lilac irises. The Buckeye is wide-ranging throughout most of North America in the summer, but it is not able to winter very far north. In Florida it is sometimes confused with the Caribbean Buckeye, a similar species that some authorities regard as synonymous with *J. coenia*. Buckeyes usually fly low over the ground, at a height of around 30–45 cm.

- Wingspan: 50–65 mm
- Range: Canada, United States, and into Central America
- Flight period: Year round
- Larval host plants: Common on plantains (*Plantago*)
- Caterpillar: See Figure 46

Caribbean Buckeye: *Junonia evarete* FIGURE 47

Southward of a line from Fort Myers on the west coast and Orange County on the east coast there exists another species of Buckeye butterfly in Florida; one with much smaller eye spots on its wings. This is the Caribbean or West Indian Buckeye, a species more common through the Greater and Lesser Antilles than in Florida. The species has a predominantly coastal range and is associated with mangrove habitat because its larval food plant is black mangrove. As the black mangrove habitat stops in central Florida, so does the range of the Caribbean Buckeye. In Jamaica, another species (*J. genoveva*) occurs which is practically indistinguishable from *J. evarete*.

- Wingspan: 50–65 mm
- Range: Central America and the Caribbean
- Flight period: Year round
- Larval host plants: Black mangrove (*Avicennia germinans*); blue porterwood (*Stachytarpheta jamaicensis*)
- Caterpillar: Black, with white spots and black spines

FIGURE 45 Buckeye (*Juonia coenia*)

FIGURE 46 Buckeye (*Juonia coenia*) caterpillar

FIGURE 47 Caribbean Buckeye (*Juonia evarete*)

FIGURE 48   White Peacock (*Anartia jatropae*)

FIGURE 49   Red Anartia (*Anartia amathea*)

FIGURE 50   Red Rim (*Biblis hyperia*)

White Peacock: *Anartia jatrophae*   FIGURE 48

The White Peacock is a common butterfly of open country, roadsides, beaches, and wastelands, often in association with other species, such as the Buckeye. It flies all year round in all the islands and in central and south Florida. The upperside is light grey with six black dots. The White Peacock is sometimes more common in swampy or wet regions, where its larvae feed on water hyssop (*Bacopa monnieri*) and Ruellia (*Ruellia occidentalis*), but it is also seen in flight over disturbed areas. In flight, *A. jatrophae* stays close to the ground and settles on flowers such as *Lantana*, *Bidens*, *Tournefortia* and *Cordia* species.

- Wingspan: 50–60 mm
- Range: South Texas and Florida, Central and South America to Argentina, Greater and Lesser Antilles
- Flight period: Year round
- Larval host plants: Water hyssop and, in central Florida, *Lippia*.
- Caterpillar: Black, spotted with silver-white and black spines

Red Anartia: *Anartia amathea*   FIGURE 49

The Red Anartia is red and black with a sprinkling of white spots on the forewing. The red of the male is deep and vivid. In the female the red shows a brownish pallor. Distributed throughout the Lesser Antilles this species becomes more common moving southward toward South America. In Trinidad, the Coolie, as it is locally known, is easily the most common butterfly. Two related species are known from the Caribbean; *A. lytrea* from the Dominican Republic and *A. chrysopelea* from Cuba, both with no red on the wings, and with white bars instead of white dots.

- Wingspan: 50–60 mm
- Range: South America and Lesser Antilles
- Flight period: Year round
- Larval host plants: Various species of Acanthaceae
- Caterpillar: Black with red-black spines

Red Rim: *Biblis hyperia*   FIGURE 50

The upper side of the adult Red Rim butterfly is a quite distinctive, velvety dark brown with a broad red band running around the border of the hindwing. The Red Rim is a fairly slow flier, found most commonly at the edge of forests where it often settles on leaves two metres or more from the ground. It is common on most islands in the Greater Antilles (except Cuba and Jamaica) and the Lesser Antilles.

- Wingspan: 50–55 mm
- Range: Mexico through Paraguay
- Flight period: Year round
- Larval host plants: Pine nettles (*Tragia volubilis*)
- Caterpillar: Grey-brown and spiny, with pale oblique lateral stripes

Malachite: *Siproeta stelenes* FIGURE 51

One of the few butterflies with green coloration, the Malachite is a beautiful insect. It is typically a tropical species but it is also known from Dade and Monroe Counties in southern Florida. More sightings have been reported from Florida in recent times, and hundreds were seen in the Homestead vicinity in 1979 feeding on rotten fruit. This is typically a butterfly of forest edges, plantations, and shady areas where it flies low to the ground, and usually below two metres. It alights frequently on the upperside of leaves and flexes its wings slowly. It is widespread throughout the Caribbean.

- Wingspan: 65–75 mm
- Range: Southern Texas to Brazil, southern Florida, and the Caribbean
- Flight period: Year round
- Larval host plants: *Blechum* spp., *Ruellia*
- Caterpillar: Velvety black, with pink prolegs, spiny warts, and coloured spines

Pearl Crescent: *Phyciodes tharos* FIGURE 52

Common throughout Florida for much of the year, this small species is found in open spaces, fields, roads, and streamsides. Present in most of the United States, the Pearl Crescent is noticeable because of its commonness and the male's habit of darting out to investigate any passing form, including human, bird, or other butterfly. The closely related and very similar-looking *P. phaon* is found in Florida, Cuba, and the Cayman Islands.

- Wingspan: 25–35 mm
- Range: Canada, United States to Mexico
- Flight period: Year round
- Larval host plants: Various composites, smooth-leaved asters
- Caterpillar: Chocolate brown, peppered with white dots, and with cream dorsal and lateral lines

FIGURE 51   Malachite (*Siproeta stelenes*) –
front and side views

FIGURE 52   Pearl Crescent (*Phyciodes tharos*) –
front view

FIGURE 53   Painted Lady (*Vanessa cardui*)

FIGURE 54   American Painted Lady (*Vanessa virginiensis*)

54

Painted Lady: *Vanessa cardui* FIGURE 53

Truly a cosmopolitan species, the Painted Lady is found in Africa, Europe, Asia, and North America; there are few areas where there is not some likelihood of seeing it. This wide distribution may be due to the very catholic tastes of the caterpillars, which prefer thistles (*Cirsium* spp.), but which will feed on a huge number of other composites. Though cosmopolitan, it is a rare insect in Florida and the Caribbean, often as a migrant from North or South America.

- Wingspan: 50–60 mm
- Range: Worldwide, but uncommon in West Indies
- Flight period: Year round
- Larval host plants: Many composites, but especially thistles (*Cirsium* spp.)
- Caterpillar: Not known to breed regularly in south Florida or the Caribbean the caterpillar is grey-brown, with transverse lines and many variously coloured black-tipped spines

American Painted Lady: *Vanessa virginiensis* FIGURE 54

The American Painted Lady is a common butterfly in much of North America, but a vagrant in much of south Florida and the Caribbean, hence the worn condition of most individuals, including the one in Figure 54. Like the related Painted Lady (*V. cardui*), it is a strong flier, but unlike *V. cardui*, it is found most commonly on mountain tops, suggesting a preference for lower temperatures consistent with its commonness in Nearctic lowlands.

- Wingspan: 45–55 mm
- Range: Canada to Central America
- Flight period: Year round
- Larval host plants: Composites, but this species is not likely to reproduce in the Caribbean
- Caterpillar: Black with narrow, yellow transverse lines, and white spots

Red Admiral: *Vanessa atalanta* FIGURE 55

In the *Vanessa* tradition, although the Red Admiral's stronghold is Europe, it is an incredible wanderer and occurs in Florida and the Caribbean as a vagrant from the north. Its habitats are open fields, deciduous forests and gardens, especially those with butterfly bush (*Buddleia*), as illustrated in Figure 55.

- Wingspan: 45–55 mm
- Range: North and Central America, Europe, North Africa, and western Asia
- Flight period: Year round
- Larval host plants: Nettles (*Urtica* spp.), but species is not likely to reproduce in the Caribbean
- Caterpillar: Black, yellow spots, and many rows of black spines with orange bases

Variegated Fritillary: *Euptoieta claudia*   FIGURE 56

The Variegated Fritillary is found throughout the state of Florida, though less commonly in the lower Florida Keys. It is much less common in the Caribbean, being found generally at higher elevations on the Greater Antilles where the closely related *E. hegesia* is more common. It is present only on Barbuda in the Lesser Antilles. The caterpillars eat an extraordinarily wide range of host plants. The adults have coloration typical of a large number of fritillaries – that is, tawny brown above, but with zigzag black lines and whitish brown below. The species frequents open areas such as fields and grasslands.

- Wingspan: 45–55 mm
- Range: Eastern United States to Argentina, but generally in tropical locations
- Flight period: March–December
- Larval host plants: Stone crop (*Sedum* sp.), beggar's tick (*Desmodium* sp.), also on violets, passionflowers, and May apples
- Caterpillar: Dull red, with white flecks and dashes, and black spines

The '89': *Callicore aurelia*   FIGURE 57

A fairly small butterfly, the '89', is typically South American but is also fairly common in Trinidad and Tobago, preferring elevated localities. Black with metallic sheen and two gold stripes on the forewing, the figures 8 and 9 are precisely designated on the underside of the wings, visible at rest. Most other South American *Callicores* have an '88' pattern. Little is known of their life histories.

- Wingspan: About 40 mm
- Range: South America
- Flight period: Unknown
- Larval host plants: Unknown
- Caterpillar: Unknown

FIGURE 55   Red Admiral (*Vanessa atalanta*)

FIGURE 56   Variegated Fritillary (*Euptoieta claudia*)

FIGURE 57   The '89' (*Callicore aurelia*)

FIGURE 58   Flambeau (*Dryas iulia*)

FIGURE 59   Gulf Fritillary (*Agraulis vanillae*)

FIGURE 60   Gulf Fritillary (*Agraulis vanillae*)
caterpillar

58

Heliconias: Heliconiidae

The Heliconiidae is a rather small family confined to the neotropics, the tropics of the New World. Members are always brighly coloured and tend to alight with wings open, presenting a picture of breathtaking beauty. The head is often large with big eyes, thin antennae, and slender body. Heliconids often have bad odour and taste to predators, and many species mimic other noxious species.

Of the seven genera usually recognised, *Heliconius* is the largest in terms of species, but relatively few frequent the West Indies. Only three species are resident in Florida. Heliconias are fairly easy to keep in captivity and are long-lived. The naturalist William Beebe kept a pet *Heliconius charitonius* called Higgins for several months.

The caterpillars, all of which feed on *Passifloracea* (passion flowers) are generally conspicuously coloured with bare branched spines: one pair on the head, one pair on each thoracic segment, and three spines on all other segments.

Flambeau: *Dryas iulia*   FIGURE 58

The Flambeau is a wide-ranging, golden-orange butterfly common throughout the Caribbean. Each island has its own endemic subspecies, each of which is nevertheless instantly recognisable as a Flambeau. The existence of such subspecies indicates an extremely sedentary habit, with little tendency toward migration. The Flambeau is found only in Dade and Monroe Counties in Florida, particularly in the Keys. It is a butterfly of the lowlands with a tendency to visit flowers such as *Bougainvillea*, *Poinsettia*, *Lantana*, *Zinnia*, and *Bidens*.

- Wingspan: 80–90 mm
- Range: Texas to Brazil, throughout the Caribbean
- Flight period: Year round
- Larval host plants: Passion flowers
- Caterpillar: Black and spiny, with some reddish-brown coloration on the legs and last two body segments

Gulf Fritillary: *Agraulis vanillae*   FIGURE 59, 60

Another instantly recognisable species, the Gulf Fritillary is brilliant red-orange above, with silver-white teardrops underneath. As its name implies, this species haunts the Gulf of Mexico area and may sometimes be seen over water in the United States. It ranges into the more northern United States but is limited by cold weather, which neither it nor its *Passiflora* host can withstand. In the fall, Gulf Fritillaries migrate southward, flying at a height of 1–2 m and at about 16 kph, although this speed depends greatly on the wind at the time. On encountering a building or other obstacle, migrating fritillaries, like other butterflies, fly up and over it without changing direction. Melanic specimens, their wings suffused to varying degrees with black, occur from time to time. The Gulf Fritillary is commonly found at many plant species during the year, but on *Richardia scabra*, *Verbena brasiliensis*, and *Biden pilosa* most commonly. It is frequently seen throughout Florida and the Caribbean.

- Wingspan: 65–70 mm
- Range: Gulf coast states of the United States, South America, and the Caribbean
- Flight period: Year round
- Larval host plants: Passion flowers
- Caterpillar: See Figure 60

Silverspot: *Dione juno*   FIGURE 61

The Silverspot is closely related to the Gulf Fritillary and occurs in very similar habitats, such as lowland flower gardens, from Martinique southward into South America. It is distinguished by the bright silver spots on the underside, which give the butterfly its common name.

- Wingspan: 65–70 mm
- Range: Central and South America and the Caribbean from Martinique southward
- Flight period: Year round
- Larval host plants: Passion flowers
- Caterpillar: Reddish brown, with yellow spots and six rows of orange spines

Zebra Longwing: *Heliconius charitonius*   FIGURE 62, 63

The Zebra Longwing is completely distinctive, with long black wings banded with lemon yellow. It has slow wafting flight and a tendency to roost communally at night. Each member of the roost returns faithfully each night to the same perch to sleep. So sound is their sleep that one can pick a butterfly off its roost and return it later without waking any of the others. In Florida, hammocks and thickets in Everglades National Park are good places to see these butterflies. They are less common in the northern third of the state. In the Caribbean the Zebra Longwing is common on the Greater and Lesser Antilles but, for some inexplicable reason, is not found south of Montserrat.

Zebra Longwings have a slow and deliberate flight and they are

FIGURE 61   Silverspot (*Dione juno*)

FIGURE 62   Zebra Longwing (*Heliconius charitonius*)

FIGURE 63   Zebra Longwing (*Heliconius charitonius*) caterpillar

61

FIGURE 64   Doris (*Heliconius doris*) – red

FIGURE 65   Doris (*Heliconius doris*) – green

FIGURE 66   Doris (*Heliconius doris*) – blue

FIGURE 67   Small Blue Grecian (*Heleconius sara*)

found in a wide range of habitats, from city gardens to open fields to paths in light forest. The attending of female pupae by males has been known since the last century, and mating often takes place before the female has emerged from the pupal case. In 1996, the Florida legislature designated this species as Florida's official butterfly.

- Wingspan: 75–85 mm
- Range: South-eastern United States, Texas through Central and South America to Ecuador, Greater and Lesser Antilles through Monserrat
- Flight period: Year round, except in cold weather
- Larval host plants: Passion flowers
- Caterpillar: See Figure 63

Doris: *Heliconius doris*   FIGURE 64, 65, 66

The Doris is trimorphic, that is, it exists in three colour varieties. Individuals of either sex can exhibit red, green, or blue coloration (see Figures 64, 65 and 66), in addition to the basic black pattern with yellow blotches. The differentiating colour occurs at the basal area of the lower wing. The most common variety is the red. Typically a South American species, it also ranges into Trinidad. The Doris prefers low-lying forested areas and flies fairly low to the ground.

- Wingspan: 70–80 mm
- Range: Central America and tropical South America
- Flight period: Unknown
- Larval host plants: *Passiflora serrato*
- Caterpillar: Unknown

Small Blue Grecian: *Heliconius sara*   FIGURE 67

The Small Blue Grecian is an inhabitant of cool shaded forest undergrowth, or the tracks and shady outskirts of forest areas. Another Trinidadian species easily confused with the Small Blue Grecian is *Heliconius wallacei*, which at 70 mm wingspan, is slightly bigger in span than *Heliconius sara*. Relatively little is known of the life history of this species.

- Wingspan: 55–60 mm
- Range: Central America and tropical South America
- Flight period: Unknown
- Larval host plants: *Passiflora auriculata*
- Caterpillar: Unknown

Postman: *Heliconius melpomene*

FIGURE 68

The Postman is very dark brown with a broad, red, central forewing band. It is seen only in Trinidad and Tobago, but very frequently there in forested areas and shaded tracks. It is very similar to its close relative, *Heliconius erato*, the Crimson-patched Longwing. One of the best ways to distinguish the two species is to smell freshly caught specimens. The smell of *H. melpomene* is supposed to be similar to fried rice, while that of *H. erato* is like witch hazel!

- Wingspan: 75–85 mm
- Range: Tropical South America
- Flight period: Unknown
- Larval host plants: *Passiflora laurifolia*, known as belle apple in Trinidad
- Caterpillar: Unknown

Metalmarks: Riodinidae

The Riodinidae family contains about a thousand species, most of which occur in the American tropics. It is very poorly represented in Florida and the West Indies, by one species in southern Florida, and another on Cuba and four Bahamian islands. Trinidad, with its proximity to the South American mainland, has more species. Riodinids are small butterflies which frequently have silver spots on one wing surface or the other, hence the common name of Metalmarks. The adults often perch under leaves with wings outstretched. Some species are very similar to the Lycaenidae with sluglike larvae which may be attended by ants. Other species resemble heliconids, pierids, satyrids, or even skippers and moths. The Bee (Figure 69) has clear wings like some ithomiids.

Bee: *Cholinea faunus*

FIGURE 69

The Bee is ebony at all wing margins, with all wing membranes being transparent, except for a bright red spot at the tail. This transparency makes it more difficult to spot in flight. The Bee is a sunshine lover often seen resting on foliage in twos and threes. Very little is known of its biology.

- Wingspan: 30–40 mm
- Range: Nicaragua and Venezuela to the Guianas
- Flight period: Unknown
- Larval host plants: Unknown
- Caterpillar: Unknown

FIGURE 68   Postman (*Heleconius melpomene*)

FIGURE 69   Bee (*Cholinea faunus*)

65

FIGURE 70   Atala (*Eumaaeus atala*)

FIGURE 71   Atala (*Eumaaeus atala*) caterpillar

66

Blues and Hairstreaks: Lycaenidae

The blues and hairstreaks are a vast family (more than 5000 described species) of small butterflies distributed worldwide. The shimmering blue colours that many display are structural, being produced by interference of light by the scales, rather than pigmental in nature. Unlike the metalmarks, the lycaenids generally hold their wings over their back when at rest. Flight in this family is fast, irregular, and usually short, most species being prone to long periods of settling on leaves with wings together.

The family has three divisions: the blues, the hairstreaks, and the coppers. The hairstreaks exhibit threadlike tails on the lower wing, while the blues have rounded hindwings. The other main subdivision of the Lycaenidae, the coppers, although common in temperature regions, is largely absent from the West Indies. At least 50 species of lycaenid have been recorded from Florida and the West Indies, and only six representatives are illustrated here (Figures 70–76). Most species exhibit the same general body pattern and coloration.

The stumpy sluglike caterpillars feed on flowers and fruit rather than on foliage. Many lycaenid caterpillars exude a honey-like fluid from between the terminal segments. Ants feed on this secretion and form a mutual relationship with the larvae, giving protection from insect predators and parasites in return for this sweet secretion.

Atala: *Eumaeus atala*   FIGURE 70, 71

Once common in Dade and Monroe Counties in Florida, the Atala was thought to have become extinct in the 1930s, having failed to survive the ravages of collectors and real estate developers. In 1959, Floridian George Rawson made one of the earliest attempts at restoration of the Atala in Florida by attempting to transfer butterflies into the Everglades. Although his attempts failed, individuals are now known from Broward and Dade counties in a few isolated areas, where its only host plant, coontie (*Zamia pumila*), grows. Also found on Cuba and the Bahamas, it inhabits open pinewoods where its larval food plant grows. The slow flight advertises its aposematic coloration as does the brilliant orange of the larvae (see Figure 71). Although a threatened species, it had the dubious distinction of being a pest on the cycad collection of Fairchild Tropical Gardens!

- Wingspan: 40–48 mm
- Range: South Florida, Bahamas, and Cuba
- Flight period: Year round
- Larval host plants: Cycads (*Zamia* spp.); in Florida on coontie (*Z. floridana*)
- Caterpillar: See Figure 71

Giant Hairstreak: *Pseudolycaena marsyas* (dead specimen)   FIGURE 72

The Giant Hairstreak is the largest West Indian Hairstreak, with the upper side a stunning sky blue. It has been seen in swampy locations on Trinidad, but is from forested areas of St Vincent where it often perches on leaf tips along forest trails. Little is known about its life history. The flight is swift and spectacular, with the adult sometimes alighting on *Eupatorium* or mangrove.

- Wingspan: 50–60 mm
- Range: Mexico to Brazil, Trinidad, Tobago, and St Vincent only
- Flight period: Unknown
- Larval host plants: Unknown
- Caterpillar: Unknown

Red-banded Hairstreak: *Calycopis cecrops*   FIGURE 73

Perhaps the most common hairstreak in north Florida, the Red-banded Hairstreak is common throughout the south-east United States. In common with many other hairstreaks, it has fairly long tails, which it often rubs together, giving the illusion of a false head. This behaviour may have considerable survival value as protection against predators, which strike toward the tail, allowing the butterfly to escape. It frequents open woodlands and fields and is often seen at flowers, (see Figure 73). Recorded food plants include members of the Anacardiaceae, Euphorbiaceae, and Myrtaceae. Eggs are laid on dead leaves beneath the host plant and larvae apparently feed on detritus. This is presumably why they fail to cause much damage to their host plants.

- Wingspan: 20–25 mm
- Range: South-eastern United States, and New Providence in the Bahamas
- Flight period: Year round
- Larval host plants: Anacardiaceae including sumac (*Rhus*) and Brazilian peppers (*Schinus terebinthifolius*)
- Caterpillar: Olive green, with blue-green dorsal stripe and brown hairs

Grey Hairstreak: *Strymon melinus*   FIGURE 74

The genus *Strymon* is the largest of the West Indian hairstreaks with 13 species, so there are a lot of similar-looking butterflies in the Caribbean. Florida has three species of *Strymon*. The Grey Hairstreak is primarily a seaside species and is a frequent visitor to flowers. This is

FIGURE 72   Giant Hairstreak (*Pseudolycaena marsyas*)

FIGURE 73   Red-banded Hairstreak (*Calycopis cecrops*)

69

FIGURE 74   Grey Hairstreak (*Strymon melinus*)   FIGURE 75   Cassius Blue (*leptotes cassius*)

FIGURE 76   Hanno Blue (*Hemiargus hanno*)

the most common hairstreak in south Florida, and its larvae have sometimes been regarded as pests on beans and cotton in other areas of the United States.

- Wingspan: 28–32 mm
- Range: Canada to northern south America, Grand Bahamas
- Flight period: Year round
- Larval host plants: At least 20 families, with preference to peas (Legume family)
- Caterpillar: Yellow to red-brown, with paler markings

Cassius Blue: *Leptotes cassius* FIGURE 75

There are four species of *Leptotes*. Three are rare, and the Cassius Blue which is common and well-known. In common with most of the blues, the wings are folded together at rest, (see Figure 75) so that the blue coloration on the upper surface of the wings does not show. The Cassius Blue is particularly attracted to blue and violet flowers. It is a sun-loving butterfly of open grassy scrubland.

- Wingspan: 15–20 mm
- Range: Texas to Argentina, central and south Florida, throughout the West Indies
- Flight period: Year round
- Larval host plants: Primary Legumes such as rattlebox (*Crotalaria*)
- Caterpillar: Green with a red wash and a pink lateral stripe

Hanno Blue: *Hemiargus hanno* FIGURE 76

Widespread in Florida and the Caribbean, the Hanno Blue is a butterfly of open roadsides, gardens, meadows, and the flowery margins of woods (see Figure 76). It has a weak flight but visits many different types of flowers, and will fly up to flowers on trees and shrubs, too. The Hanno Blue is sometimes found in turkey-oak forests and beach dunes – a real habitat generalist.

- Wingspan: 20 mm
- Range: South Texas through Central and South America to Argentina. Throughout Florida and the West Indies
- Flight period: Year round
- Larval host plants: Various Legumes such as *Crotalaria*
- Caterpillar: Yellow-green with various dark green, pink, or light yellow stripes

Whites and Sulfurs: Pieridae

Pierids, comprising almost 2000 species worldwide, are some of the most abundant butterflies in all regions. The brimstones, yellows, or sulfurs are notable migrants and may assemble on damp soil or mud to drink, providing quite a spectacle. The larvae of several pierids are of economic importance as pests, especially on Crucifers (whites) and Legumes (sulfurs). Caterpillars are usually smooth, green, and cylindrical. The affinities of most Caribbean species lie in the Neotropics. West Indian species, of which there are about 50, are medium-sized butterflies with wingspans between 3 and 6 cm. The word butterfly is probably derived from a member of this family: a butter-coloured fly. The colours are based on pigments rather than structural separation of light. In the case of the whites, it is derived from a waste product of the insect's metabolism incorporating uric acids.

Florida White: *Appias drusilla*   FIGURE 77

The Florida White may be found throughout the year in Dade and Monroe Counties, though only rarely. It is more commonly a West Indian and South American species. Males are brilliant white on both surfaces, with scarcely a hint of black. Females are more variable, with off-white coloration, sometimes with dark borders on the forewing, especially on most Caribbean islands. Florida Whites are generally found in open lowlands near dry woods. The closely related *A. punctifera*, found in Hispaniola, Puerto Rico, and the Virgin Islands, has a small black spot on the forewing.

- Wingspan: 40–60 mm
- Range: South Florida to southern Brazil, throughout the West Indies
- Flight period: Year round
- Larval host plants: Jamaican copper tree (*Drypetes tateriflora*)
- Caterpillar: Grey-green with maroon dorsal spots and a greenish-white lateral line

Great Southern White: *Ascia monuste*   FIGURE 78

Numerous toward coastal areas, the Great Southern White is usually slightly larger than the Florida White and has dark tips on the wings. Some females may be entirely suffused with smoky scales, more commonly in high summer. These whites undergo large-scale migrations, probably when the local food supply is scarce. During such times they may be seen in abundance along waterways of the east coast of the United States, where larvae may feed on saltwort in salt flats. In some

FIGURE 77   Florida White (*Appias drusilla*)

FIGURE 78   Great Southern White (*Ascia monuste*)

FIGURE 79   Cabbage White (*Pieris rapae*)

FIGURE 80   Little Sulfur (*Eurema lisa*)

74

areas the caterpillars are common enough to be pests on Crucifers such as kale, raddishes, cabbages, and mustard.

- Wingspan: 45–65 mm
- Range: Gulf coast states of the United States, throughout the Caribbean and South America
- Flight period: Year round
- Larval host plants: Saltwort (*Batis maritima*), sea rocket (*Cakile edentule*) and various Crucifers
- Caterpillar: Green, with a yellow mid-dorsal line, purple-grey stripes, and maroon and orange spots

Cabbage White: *Pieris rapae*   FIGURE 79

A native of Europe, North Africa, and temperate Asia, this butterfly was introduced into the New World in Quebec around 1860, and spread rapidly throughout North America. It is absent from the Caribbean, rare in south Florida, and more common in central and north Florida. Found in fields and disturbed habitats, it is best known in gardens where the larvae are pests of cabbages and other Brassicaceae. Its spectacular success in the new world has been blamed, probably incorrectly, on the decline or retreat of some of its indigenous relatives.

- Wingspan: 35–50 mm
- Range: Worldwide, in temperate areas
- Flight period: Early spring to fall
- Larval host plants: Crucifers, especially cabbages and nasturtiums
- Caterpillar: Green

Little Sulfur: *Eurema lisa*   FIGURE 80

Nearly 100 species occur in the large, pantropical genus *Eurema*, with 23 species in the West Indies and five in Florida. The most common in Florida and the Caribbean is the Little Sulfur with the Barred Sulfur, (*E. daira*), being another frequently encountered species. The Little Sulfur is common in open fields and woods where the adults visit small flowers, especially Legumes. The flight is low to the ground and weak.

- Wingspan: 25–35 mm
- Range: Southeastern United States to Costa Rica, and throughout the Caribbean to Barbados
- Flight period: Year round
- Larval host plants: Sennas (*Cassia* spp.)

- Caterpillar: Green, darker underneath, with a bright yellow lateral line and cream spiracles

Sleepy Orange: *Eurema nicippe*   FIGURE 81

A small sulfur, bright gold-orange above and with an uneven black border, the Sleepy Orange can be prolific in the southern United States including Florida, with adults visible in all months. The common name may derive from its habit of hibernating through the cooler days of southern winters. The species has a rapid zigzag flight, often in full sun. Males are often abundant at mud puddles. It is a butterfly of open grassy areas and fields.

- Wingspan: 35–50 mm
- Range: South-eastern United States to Costa Rica; Greater Antilles and Bahamas
- Flight period: Year round
- Larval host plants: Sennas (*Cassia* spp.)
- Caterpillar: Green, with yellow lateral stripes, speckled with black. Body tapers toward head and tail

Dwarf Yellow: *Nathalis iole*   FIGURE 82

The Dwarf Yellow is the smallest North American and Caribbean pierid, with black tips on the forewings and diffuse black bars along the inner margins of both fore and hindwings. It is generally found in dry pastures, sandy coastal scrub, and exposed hillsides, roadsides, and river banks. It flies low to the ground, feeding on numerous flowers. The Dwarf Yellow often migrates into central North America, from the desert south-west of the United States, in the summer.

- Wingspan: 20–30 mm
- Range: Gulf coast states of United States and desert south-west of the United States, Greater Antilles, Bahamas, Central America to Colombia
- Flight period: Year round
- Larval host plants: Weedy composites, asters, bur marigolds (*Bidens* spp.), chickweed (*Stellaria*)
- Caterpillar: Deep green, with purple stripes and black and yellow side stripes

FIGURE 81   Sleepy Orange (*Eurema nicippe*)

FIGURE 82   Dwarf Yellow (*Nathalis iole*)

FIGURE 83   Orange Sulfur (*Colias eurytheme*)

FIGURE 84   Dogface (*Colias cesonia*)

Orange Sulfur: *Colias eurytheme* FIGURE 83

There are many species of *Colias* present in the United States, but only three in Florida and none in the Caribbean. The Orange Sulfur, *C. eurytheme*, has an orange wash to the upper side of its yellow wings. It is common in mowed and open fields. The Common Sulfur, *C. philodice*, has almost continuous black marginal bands on the wings and is a rare visitor to northern Florida and the Panhandle.

- Wingspan: 40–60 mm
- Range: Continental United States but rare in south Florida
- Flight period: March–December
- Larval host plants: Many herbaceous Legumes such as clover (*Trifolium repens*) and alfalfa (*Medicago sativa*)
- Caterpillar: Grass green and covered in long white hair with pink stripes low on the sides, white stripes higher

Dogface: *Colias cesonia* FIGURE 84

When the wings of the Dogface are spread the black margins on the upper surface of the forewings outline the profile of a dog's face or poodle's head. As most individuals rest with wings closed, photographs of live specimens (as here) do not generally show this feature. Present only in northern and central Florida, and not the West Indies, the Dogface is a denizen of open woodlands and sandy scrub.

- Wingspan: 48–64 mm
- Range: California to Florida, south to Argentina
- Flight period: Year round, in warm conditions
- Larval host plants: Indigo bush (*Dalea* sp.) (a member of the pea family), and clovers such as sweet clover (*Trifolium* sp.)
- Caterpillar: Green, with black stippling, sometimes with stripes or cross-bands of yellow and black

Cloudless Sulfur: *Phoebis sennae* FIGURE 85, 86

The Cloudless Sulfur is almost pure yellow in coloration. It is the most common large sulfur throughout Florida, the Caribbean, and indeed the New World. The common name stems from the butterfly's tendency to fly only on sunny, 'cloudless' days. During migrations individuals have been clocked flying at about 13 kph into a 8 kph head wind; they normally fly about 1 m above ground level, exhibiting an oscillating flight path. Summer and autumn movement sometimes brings many individuals into the northern United States, but they always die out in the cold weather. The Cloudless Sulfur is found in a wide range of habitats, from gardens and fields to disturbed areas. Some caterpillars hide within tents formed of silk and of leaves of their host plants, normally Legumes, but others occur right out in the open on *Cassia obtusifolia*, a favourite food plant.

- Wingspan: 55–70 mm
- Range: Gulf coast states of United States, California, Central and South America to Argentina, throughout the West Indies
- Flight period: Year round
- Larval host plants: *Cassia* spp.
- Caterpillar: Typical (see Figure 86)

Orange-barred Sulfur: *Phoebis philea* (dead specimen)   FIGURE 87

The genus *Phoebis* contains six large Caribbean sulfurs, three of which occur in south Florida. The Orange-Barred Sulfur is the largest of the commonly encountered Caribbean sulfurs, with a wingspan of up to 8.5 cm. Its coloration is bright yellow with a broad orange band bordering the hindwing, and an orange bar across the centre of the forewing. The Orange-Barred Sulfur is common only in the southern half of Florida, as it is typically a tropical species. *P. philea* can often be seen drinking on river banks and feeding on shrubs such as *Hibiscus* and *Bougainvillea*. Otherwise it is a high-flying butterfly. The related *P. avellaneda*, which is found only in Cuba, has much more orange on the wings.

- Wingspan: 70–85 mm
- Range: Mexico to southern Brazil, south Florida, Grand Bahamas, Cuba, Hispaniola, and Puerto Rico
- Flight period: Year round
- Larval host plants: Senna (*Cassia* sp.) and royal poinciana (*Poinciana pulcherima*)
- Caterpillar: Yellow-green, with black grainy dots, orange side brand, wrinkled and tapered

Large Orange Sulfur: *Phoebis agarithe*   FIGURE 88

The Large Orange Sulfur is a swift-flying butterfly that may pause to feed at trees and shrubs such as *Ixora, Bougainvillea, Hibiscus,* and *Lantana*. Similar to, but more common than *P. argante* which is found in the Greater Antilles and a few islands of the Lesser Antilles. The two species may be distinguished by their wing coloration. *P. argante* has black marginal dots on the wings while *P. agarithe* does not. Also *P. agarithe* prefers more open land than *P. argante*, which is attracted to forest edges.

FIGURE 85   Cloudless Sulfur (*Phoebis sennae*)

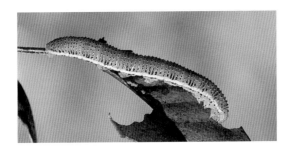

FIGURE 86   Cloudless Sulfur (*Phoebis sennae*) caterpillar

FIGURE 87   Orange-barred Sulfur (*Phoebis philea*)

81

FIGURE 88   Large Orange Sulfur (*Phoebis agarithe*)

FIGURE 89   Zebra Swallowtail (*Protesilaus marcellus*) – side and top views

82

- Wingspan: 55–65 mm
- Range: South Florida, south Texas through Central America to Peru, throughout the Caribbean
- Flight period: March–December in Florida
- Larval host plants: Shrubby and tree Legumes including *Inga* and *Cassia* spp., and *Pithec ellobium* spp., such as black beard and cat claw
- Caterpillar: Light green, with a thin yellow side-stripe edged with dark green

Swallowtails: Papilionidae

Swallowtails, named for the long tails that project from the hindwings, are among the West Indies' most striking butterflies and can be frequently seen at flowers. So typical of butterflies are members of this group that they were named using the Latin word for butterfly, *papilio*. The family contains about 700 species distributed throughout the world, with about 25 species common in Florida and in the Caribbean.

Papilionids can be broadly divided into three groups: true swallowtails (*Papilio*); kite swallowtails (*Eurytides*), which have narrow, pointed wings, giving the appearance of an old-fashioned kite; and poison eaters (*Battus*), so called because the larvae feed on *Aristolochia* vines, from which the larvae derive poisonous substances. All the larvae are smooth-skinned and possess a structure called an osmeterium, a forked horn that can be erected from behind the caterpillar's head. It emits a foul-smelling odour and is particularly useful in detering would-be predators. The smell of each species varies according to the plant on which it feeds.

The adult butterflies are very colourful, with black and yellow predominating. The swallowtails (*Papilio* sp.) are generally black with yellow spots or bands, kite swallowtails (*Eurytides* sp.), black with white stripes, and Pipevine swallowtails (*Battus* sp.) blackish and blue-green. The bright colours generally advertise distastefulness to predatory birds.

Zebra Swallowtail: *Protesilaus marcellus*   FIGURE 89, 90

The Zebra Swallowtail is a distinctive butterfly with long swordlike tails issuing from the hindwing. It has at least three generations per year in Florida, first generation adults being smaller and paler than the subsequent two. Of all the swallowtails in Florida, this one is more common in woodland than the others, but it is still frequent in fallow fields and flowering roadsides. The flight is rapid and low, and its range is regulated by supply of its larval food, wild pawpaw (*Asimina*) and its relatives, which are found more commonly in pine woodlands.

Three other very similar species occur in the Caribbean: *P. celadon* in Cuba, *P. zonarius* in Hispaniola, and *P. marcellinus* in Jamaica.

- Wingspan: 60–90 mm
- Range: South-eastern United States
- Flight period: March–December
- Larval host plants: Pawpaws (*Asimina triloba* and *A. reticulata*)
- Caterpillar: See Figure 90

Cattle Heart: *Parides anchises*   FIGURE 91

Three species of *Parides* occur in the Caribbean. The Cattle Heart is a beautiful South American species found in the Caribbean only in Trinidad. It is black with red lobes on the hindwing and white dots on the forewing. It is a lover of deep tropical forest and generally avoids sunshine. It likes to fly just after rains. Closely related, and often found in the same forested areas, is *Parides neophilus* the Spear-winged Cattle Heart, which is distinguished by its more pointed wings. Males of both species have additional green-blue patches in the central forewing. Caterpillars of both feed on *Aristolochia*.

The beautiful *P. gundlachianus*, which flies in Cuba, was described by Gundlach in 1857 after he shot a specimen with bird shot! It has green on the upper forewing, blue on the lower forewing, and red on the upper and lower hindwing.

- Wingspan: 65–75 mm
- Range: Colombia to Bolivia
- Flight period: Unknown
- Larval host plants: Unknown
- Caterpillar: Unknown

Pipevine Swallowtail: *Battus philenor*   FIGURE 92, 93

The wings of the Pipevine Swallowtail are usually black with a blue-green iridescence, which gives rise to its alternative name, 'Blue Swallowtail'. A wide-ranging species, it occurs throughout northern and central Florida as far south as Fort Myers. Horticulture has caused the spread of pipevines (*Aristolochia* sp.) the larval food of this butterfly, thereby extending its range. The adult flies slowly over open country, alighting at thistles (*Cirsium*) and *Bidens alba* to feed. The closely related *B. devilliers*, which flies in Cuba and the Bahamas, is similar in appearance, but is larger and has red and yellow on the underwing, with no traces of a metallic blue sheen.

The pipevines are noxious, giving the larvae and hence the adults a bad taste. As a result, many birds learn to reject Pipevine Swallowtails

FIGURE 90   Zebra Swallowtail (*Protesilaus marcellus*) caterpillar

FIGURE 91   Cattle Heart (*Parides anchises*)

FIGURE 92   Pipevine Swallowtail (*Battus philenor*) – top and side views

FIGURE 93   Pipevine Swallowtail (*Battus philenor*) caterpillar

FIGURE 94   Gold Rim (*Battus polydamas*)

as food. Other species, such as the Spicebush Swallowtail and Red-spotted Purple (a nymphalid), are similar in colour and may have evolved to mimic *B. philenor*.

- Wingspan: 70–85 mm
- Range: Canada to Costa Rica
- Flight period: February–November
- Larval host plants: Dutchman's pipe (*Aristolochia macrophylla*)
- Caterpillar: See Figure 93

Gold Rim: *Battus polydamas* FIGURE 94

Also called the Polydamas Swallowtail or Black Page, the Gold Rim is a widespread species occurring from south Florida and Texas, through the Caribbean and South America, to Argentina. Each Caribbean island plays host to a slightly different variation or race, indicating little migration between islands and probable isolation for long periods of time. The basic pattern is black, with a submarginal rim of yellow rectangular spots on the lower wing, continuing less prominently on the forewing. Island varieties have different amounts of green replacing the yellow on the upper hindwing. Habitat includes gardens, fields, disturbed areas, and beachside scrub. In the related *B. zetides*, which flies only on Hispaniola, the gold rim is replaced by a thinner red rim on both wings.

- Wingspan: 80–100 mm
- Range: Texas, Florida, Central and South America, and throughout the West Indies
- Flight period: Year round
- Larval host plants: Dutchman's pipe (*Aristolochia macrophylla*)
- Caterpillar: Black or tan, with a yellowish lateral band, dark red tubercles, and many fleshy spines

Tiger Swallowtail: *Pterourus glaucus* FIGURES 95, 96, 97

The Tiger Swallowtail exhibits a striking form of dimorphism, which is confined to the female sex. Males and some females are yellow with black tiger stripes across the wings (see Figure 95). Some females, however, exhibit a dark form, being black above with a bordering of yellow, blue, and orange (see Figure 96). This form is generally predominant south of latitude 40°N and is thought to be a mimic of the Pipevine Swallowtail.

The Tiger Swallowtail is a common Floridian butterfly often encountered near woodlands. In Florida, Tiger Swallowtails can be very large, even exceeding the Giant Swallowtail in size. The adults feed on nectar

FIGURE 95   Tiger Swallowtail (*Pterourus glaucus*) – male

FIGURE 96   Tiger Swallowtail (*Pterourus glaucus*) – female

88

FIGURE 97   Tiger Swallowtail (*Pterourus glaucus*) caterpillar

FIGURE 98   Green Clouded Swallowtail *(Pterourus troilus)* – front view

of many flowers including blazing star, *Liatris*, (see Figure 95). They are also commonly seen drinking from sand.

The earliest known picture of an American butterfly is one of the Tiger Swallowtail, painted by John White, commander of Sir Walter Raleigh's third expedition to Virginia in 1587.

- Wingspan: 80–140 mm
- Range: Alaska to the Gulf coast. Rare in south Florida
- Flight period: March–November
- Larval host plants: In Florida, primarily sweetbay (*Magnolia virginia*); also wild cherry (*Prunus*)
- Caterpillar: Resembles a bird dropping when small, but when mature is green and has large false eyespots (see Figure 97).

Green Clouded Swallowtail: *Pterourus troilus*   FIGURE 98, 99, 100

Sometimes named the Spicebush Swallowtail in deference of its use of this plant as a larval food source, the Green Clouded Swallowtail is named for the distinct blue-green wash on the hindwings. It occurs in woods, fields, and rights of way. Like black female tigers, it mimics the Pipevine Swallowtail.

- Wingspan: 90–110 mm
- Range: Eastern North America, rare in south Florida
- Flight period: March–December
- Larval host plants: Sweetbay (*Magnolia virginia*); and further north, spicebush (*Lindera benzoin* and *Sassafras*)
- Caterpillar: See Figure 100.

Palamedes Swallowtail: *Pterourus palamedes*   FIGURE 101, 102

The Palamedes Swallowtail is often encountered in coastal areas, where its habitats of subtropical wetland, or humid woods with standing water, may be present. It is especially common in big swamps such as the Everglades and Big Cypress. The adults love to take nectar from pickerelweed and they may roost communally in oaks and palmettos. The Palamedes Swallowtail is found only in Florida and not in the West Indies.

Another member of this genus in the Caribbean is the highly prized *P. homerus*, the largest swallowtail in the Americas, with a wingspan exceeding 155 mm. It has more rounded wings and extensive powdery blue markings on the upper surface of the hindwing. *P. homerus* is restricted to Jamaica and is rare and endangered there because of collection of adults and, perhaps more importantly, habitat destruction.

FIGURE 99   Green Clouded Swallowtail (*Pterourus troilus*) – side view

FIGURE 100   Green Clouded Swallowtail (*Pterourus troilus*) caterpillar

91

FIGURE 101   Palamedes Swallowtail (*Pterourus palamedes*) – top and side views

FIGURE 102   Palamedes Swallowtail
(*Pterourus palamedes*) caterpillar

FIGURE 103   Giant Swallowtail
(*Heraclides cresphontes*)

FIGURE 104   Giant Swallowtail
(*Heraclides cresphontes*)
caterpillar

- Wingspan: 80–140 mm
- Range: Virginia to Texas and Mexico, throughout Florida
- Flight period: March–December
- Larval host plants: In wet areas on red bay (*Persea borbonia*) and avocados (*P. americana*)
- Caterpillar: See Figure 102

Giant Swallowtail: *Heraclides cresphontes*   FIGURE 103, 104

The Giant Swallowtail is America's largest butterfly, with a wingspan of up to 15 cm, though in Florida it is rivalled by the Tiger Swallowtail. It is a widespread and strong-flying species ranging from southern Canada across most of the eastern United States and down into Mexico. The larva, called the 'orange dog' by orange growers, is sometimes considered a pest on *Citrus*, but only on young trees.

The Giant Swallowtail occurs only in Florida, where it is common in gardens and fields. In Cuba and Jamaica it is replaced by the very similar *H. thoas*, in Hispaniola by the similar *H. machaonides*, and in the Bahamas by the similar *H. andraemon* (which also occurs in Cuba and Jamaica, and occasionally in the Florida Keys). Thus, each island or series of islands seems to have its own variety of Heraclides that has evolved *in situ*. All these species have predominantly yellow undersides to their wings. In addition, there are a series of Caribbean swallowtails with black underwings, but with similar markings on the upper wings, like *H. oxynius* in Cuba, and *H. aristor* in Hispaniola.

- Wingspan: 90–150 mm
- Range: Canada through North America and Central America to Colombia
- Flight period: Year round
- Larval host plants: Various *Citrus* trees (*Rutaceae*) and torchwood (*Amyris elemifera*)
- Caterpillar: A birdlime mimic with a red osmeterium (see Figure 104)

Thoas Swallowtail: *Heraclides thoas*   FIGURE 105

The Thoas Swallowtail almost matches the Giant Swallowtail for size. In the Caribbean it is found in Cuba where it is widespread, in Jamaica where it is sporadic, and in Trinidad. It has a lazy flight, alighting on *Poinsettia* and *Bougainvillea* to feed. Common around flowers in sunshine, this butterfly is so similar to the Giant Swallowtail that the two are easily confused. The yellow coloration in the centre of the tails on the hindwings is more extensive in *H. cresphontes* than in *H. thoas*, and so is the blue coloration above the orange dot.

- Wingspan: 100–140 mm
- Range: Southern Texas to Argentina; Cuba and Jamaica
- Flight period: December–February, July–August
- Larval host plants: Various Rutaceae such as *Citrus* and *Amryris*
- Caterpillar: Olive and off-white, resembles a bird dropping

Dusky Swallowtail: *Heraclides aristodemus* (dead specimen)

FIGURE 106

The Dusky Swallowtail is butterfly of dry lowland scrub and tropical hardwood hammocks, only occasionally straying into open fields and gardens. It is shy and not easily approached. Five subspecies have been described for Cuba, Hispaniola and Puerto Rico, Great Inagua (Bahamas), Crooked Island (Bahamas), south Florida and the northern Bahamas. The subspecies *Heraclides aristodemus ponceanus*, or Schaus' Swallowtail from south Florida and the northern Bahamas, has been listed as critically endangered globally. Once common in the Miami area, it has long since been excluded from that area by development and was only rediscovered on the Biscayne National Monument by Larry Brown and his co-workers in the 1970s. However, plantings of its natural host, torchwood, may encourage an expansion of its range, which, in Florida, is currently confined mainly to Elliott Key and north Key Largo. The endangered status of Schaus' Swallowtail was acknowledged by the US Postal Service in October 1996 when it was included in a panel of fifteen 32-cent commemorative stamps depicting animals on the federal endangered species list.

- Wingspan: 85–95 mm
- Range: Extreme south Florida, Bahamas, Cuba, Hispaniola, Puerto Rico
- Flight period: April–October
- Larval host plants: Torchwood (*Amyris elemifera*), and wild lime (*Zanthoxylum fagara*)
- Caterpillar: Dark brown, with creamy blotches and blue dots

Androgeus Swallowtail: *Heraclides androgeus*

FIGURES 107, 108, 109

In the Greater Antilles the Androgeus Swallowtail is most often seen in open agricultural land, especially near *Citrus* sp. trees. It flies powerfully at a height of several metres. Introduced into south Florida in 1976, it may still exist there, but it has not been seen in Florida since 1983. The sexes are very dissimilar (see Figures 106 and 107), with the male having strong black and yellow markings and the female black and green on the hindwing.

In Jamaica, a similar, large, sexually dimorphic swallowtail,

FIGURE 105   Thoas Swallowtail (*Heraclides thoas*)

FIGURE 106   Dusky Swallowtail (*Heraclides aristodemus*)

FIGURE 107   Androgeus Swallowtail (*Heraclides androgeus*) – male

FIGURE 108   Androgeus Swallowtail (*Heraclides androgeus*) – female

FIGURE 109   Androgeus Swallowtail (*Heraclides androgeus*) caterpillar

FIGURE 110   Eastern Black Swallowtail (*Papilio polyxenes*)

FIGURE 111   Eastern Black Swallowtail (*Papilio polyxenes*) caterpillar

H. thersites, occurs with males similar to *H. androgeus*, and with females black with a series of small red chevrons, together with the marginal yellow crescentic spots.

- Wingspan: 100–150 mm
- Range: Mexico to Argentina, Greater Antilles
- Flight period: Throughout the year but more common in June to August
- Larval host plants: Orange (*Citrus sinensis*)
- Caterpillar: See Figure 109

Eastern Black Swallowtail: *Papilio polyxenes*   FIGURE 110, 111

This common Floridian butterfly often flies low to the ground, coming to rest on low vegetation, such as milkweed, to feed. It is the smallest of the predominantly black swallowtails with yellow markings. Although it may also be found on Grand Bahamas, it is not common there or in extreme southern Florida and the Keys.

- Wingspan: 65–90 mm
- Range: Eastern North America from Canada into Mexico and Peru
- Flight period: Year round, in warm weather, but commonly February to September
- Larval host plants: Wild carrots (*Daucus carota*) and other Umbellifers
- Caterpillar: See Figure 111

Skippers: Hesperidae

Skippers derive their popular name from a darting, skipping flight, quite unlike that of other butterflies. In many ways they can be regarded as primitive lepidopterans, with many similarities to moths, such as stout bodies and the ability to rest with forewings folded flat on each side of the body. The family is a large one with 67 Floridian species and over 130 in the Caribbean, too many to cover in detail here. Instead, just three species are represented as examples of the group. (See Figures 112–115). Most are small, with dull or subdued colours of browns and occasional lighter markings.

Caterpillars are commonly green or whitish and often feed on grasses, weaving silk and leaf shelters during the day. They may have an enlarged head with a somewhat constricted neck or collar, on the first thoracic segment.

FIGURE 112   Long-tailed Skipper (*Urbanus proteus*)

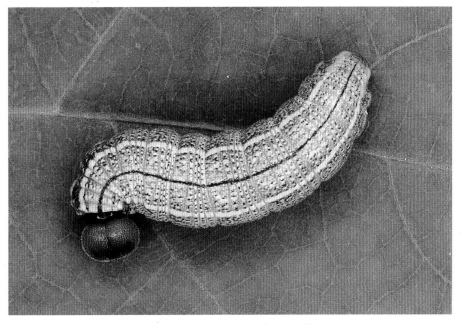

FIGURE 113   Long-tailed Skipper (*Urbanus proteus*) caterpillar

FIGURE 114   Sickle-winged Skipper (*Achylodes mithrades*)

FIGURE 115   Brazilian Skipper (*Calopedes ethlius*) caterpillar

100

Long-tailed Skipper: *Urbanus proteus* FIGURE 112, 113

This is one of the most widely distributed skippers, indeed butterflies, in Florida and Caribbean. The larvae are sometimes a pest on beans, and the caterpillar is known as the bean leaf roller. Specimens characteristically possess tails on the hindwing. The species is common from gardens to meadows, especially when the sun shines.

- Wingspan: 40–50 mm
- Range: California, Arizona, and the Gulf States through Central and South America to Argentina
- Flight period: Year round
- Larval host plants: Many species of peas (Legumes) and beans (Crucifers)
- Caterpillar: See Figure 113

Sickle-winged Skipper: *Achylodes mithridates* FIGURE 114

The Sickle-winged Skipper is found in a wide variety of habitats, from dry woodland and shady trails to disturbed land. Adults visit *Lantana*, *Croton*, and *Cordia* to feed and will also drink from moist soil. The hooked tips and violet coloration separate this species from any other skipper. It is not present in Florida.

- Wingspan: 40–50 mm
- Range: Texas to Argentina, and throughout the Caribbean
- Flight period: Year round
- Larval host plants: Heavily spined members of the Rutaceae, such as *Citrus* spp. and prickly lime (*Zenthoxylum*)
- Caterpillar: Green, with broken yellow stripe and minute white dots and hairs

Brazilian Skipper: *Calpodes ethlius* FIGURE 115

The Brazilian Skipper is most commonly encountered in gardens as a caterpillar feeding on cannas (see Figure 115). The adult is a warm brown; darker on the outer part of the forewing and lighter on the inner part, with a prominent row of translucent spots above and below.

- Wingspan: 45–55 mm
- Range: Southern United States, throughout the West Indies and southward to Argentina
- Flight period: Year round
- Larval host plants: Canna spp. (*Canna flaccida* and *C. indica*)
- Caterpillar: See Figure 115

Owls: Brassolidae

The owls, as the Brassolidae are called, are medium to very large in size and are confined entirely to the Neotropical region. There are about 80 species in the family. Irrespective of wing shape or size, the brassolids all share a peculiarity of the undersurface pattern – a combination of zigzag striations of contrasting colours and eyespots (ocelli) of varying size and number. The eyespots supposedly afford protection from predatory birds which might strike at the eye instead of the head, or be intimidated by the two large eyespots as the butterfly flaps its wings. They are distributed from Mexico to Paraguay and Trinidad. The richest concentration of species is in the Western Amazonas. The brassolids are sluggish fliers with crepuscular tendencies. Some *Caligo* species are pests of bananas in Central America.

Owl: *Caligo teucer*   FIGURE 116, 117

As are most brassolids, the Owl butterfly is found in well-wooded areas, especially at dusk. It is well camouflaged on the underside, bearing a marked resemblance to bark or pieces of wood, on which it frequently rests with wings closed. In Trinidad and Tobago it is often found in cocoa estates and is known as the Cocoa Mort Bleu. It is often found on rotting fruit.

- Wingspan: 95–125 mm
- Range: Costa Rica through northern South America
- Flight period: Unknown
- Larval host plants: Bananas (*Musa*)
- Caterpillar: See Figure 117

Morphos: Morphidae

Morpho means beautiful or shapely, and is certainly an appropriate name for this dazzlingly beautiful family of about 80 species. The Morphidae are distributed from Central America to southern Brazil, but the greatest concentration of species is in the Amazonas and the Guianas. One species can be found in Trinidad and Tobago.

The blue iridescence is not caused by any particular pigment, but rather by an arrangement of the scales in such a way as to catch the light to reflect the iridescences perceived. This phenomenon is more prevalent in males than in females so that males are bluer than the females. Because the colours cannot fade, the butterflies have long been used in jewellery and art. Their large wing area, relative to body size, gives them an effortless soaring flight. In some parts of the

FIGURE 116   Owl (*Caligo teucer*)

FIGURE 117   Owl (*Caligo teucer*) caterpillar

FIGURE 118   Morpho (*Morphos peleides*)

Amazon, the Indians regard the Morpho as a piece of the sky fallen down to earth.

Morpho: *Morpho peleides*   FIGURE 118

Only one species of Morpho, *Morpho peleides*, commonly called simply by the family name, penetrates the Caribbean into Trinidad and Tobago. Males are often found in sun-swept gullies of tropical forests where water trickles. Females are much rarer and spend most of the day in the forest. Little else is known of their life histories.

- Wingspan: 120–130 mm
- Range: Mexico to southern Brazil
- Flight period: June–October
- Larval host plants: Leguminosae (*Paragonia pyramidata*)
- Caterpillar: Generally gregarious, and with brightly-coloured hair tufts and a tail fork

Glossary

Adult: the last stage in the life cycle; (sometimes called imago).

Allopatric: occupying different geographical areas.

Anacardiaceae: the cashew or sumac family; shrubs or small trees with resinous or milky juice.

Apical: at the tip, usually of the wing.

Aposematic coloration: bright colours which serve to warn predators of unpalatability.

Basal: toward the base; usually refers to the point of attachment of the wing.

Batesian mimic: A species which, although in itself is *not* offensive to predators, has developed a similarity in appearance to a species which *is* offensive to predators, sufficient to deter attacks.

Caterpillar: the feeding stage of butterflies; sometimes called larva.

Caudal: toward the posterior end.

Chitin: the main material forming the tough exoskeleton.

Chrysalis: the transformation stage of butterflies from caterpillar to adult; sometimes called pupa.

Composite: a plant belonging to the family Compositae, daisies or sunflowers.

Cosmopolitan: occurring in many parts of the world.

Costa: the anterior margin of the forewings and hindwings.

Cotyledons: a leaflike structure within a seed that absorbs food molecules from the endosperm and transfers them to the growing embryo.

Crucifers: members of the Cruciferae or Brassicaceae, often having a peppery sap.

Crypsis: camouflage colouring which allows a butterfly or its immature stages to blend into its background.

Deciduous: shedding leaves annually.

Dicotyledonous: a flowering plant with two seed leaves (cotyledons).

Dimorphic: occurring in two different forms.

Dorsal: the upper surface.

Eclosion: emergence of the tiny adult butterfly from the pupa.

Endemic: restricted to a particular locality or region, not found elsewhere.

Euphorbiaceae: herbs with a milky sap, some trees and shrubs in the tropics.

Family: the taxonomic category below the level of order; family names end in '-dae'.

Frass: waste products from the butterfly or, more usually, the caterpillar.

Genitalia: the sex organs.

Genus: a taxonomic category to which species are assigned, e.g., *Papilio*, a genus of swallowtails.

Hammock: a raised clump or island of dense woody vegetation surrounded by water or grassy plains.

Host plant: the food plant of the caterpillar.

Imago: the adult stage.

Instar: any one of the stages between moults in the growth of a caterpillar.

Introduced: a species not native to an area, one that was brought in by humans.

Iridescence: the metallic reflections of structurally specialised scales, usually on the wing.

Larva: the immature stage between egg and pupa, i.e., caterpillar.

Legume: a member of the pea or bean family (Fabaceae or Leguminosae).

Metamorphosis: process of development from the egg to the adult.

Migrant: a non-resident butterfly blown or flown in from another area.

Migration: a mass movement of adult butterflies, usually seasonally.

Mimicry: the phenomenon wherein two or more unrelated species closely resemble each other, usually to enhance protection from predators, (see Batesian mimic; Mullerian mimic).

Monocotyledonous: a plant with one seed leaf, usually a sedge or grass.

Mullerian mimic: Young insectivourous predators learn after several encounters, which insects are inedible. Several species sporting the same warning coloration, therefore, suffer, on aggregate, a more bearable loss of numbers.

Myrtaceae: the Myrtle family, includes punk tree, *Melaleuca*.

Ocelli: pattern on wings resembling an eye.

Osmeterium: the reversible fleshy organ, on the first segment of the thorax of swallowtail caterpillars, that emits defensive chemicals.

Oviposition: the act of laying eggs.

Parasitoid: larva of wasp or fly which lives inside one of the stages of the butterfly life cycle, most commonly the caterpillar. It feeds on its host's tissues, eventually causing its death.

Patrolling: the persistent flying of a male in a certain area as it searches for a receptive female.

Pheromone: a volatile, airborne chemical messenger, often used to attract members of the opposite sex.

Pupa: the immature stage between larva and pupa, i.e., chrysalis.

Race: a geographically distinct form of a species; sometimes called a sub-species; usually with some structural or colour difference.

Savanna: a treeless, grassy plain.

Sclerotized: hardened.

Scrub: low, dry, woody vegetation.

Sequester: the obtaining of chemicals from plants by the caterpillar, which the adult butterfly uses to deter enemies.

Shrub: a non-grassy, woody perennial with a number of stems.

Solanaceae: the Nightshade family; plants with showy flowers in branched clusters.

Species: a large group of morphologically similar, reproducing individuals.

Subspecies: a taxonomic subdivision of a species, usually with some colour differences, often with a distinct geographical distribution, but which is able to interbreed with other subspecies.

Subtropical: an area bordering the tropics.

Sympatric: occupying the same geographical area.

Taxonomy: the naming and arranging of species and higher groups into a system of classification.

Transverse: across, crosswise.

Tubercle: a protuberance or projection on a larva or pupa.

Umbellifers: the carrot (Apiaceae) or parsley (Umbelliferae) families.

Univoltine: having a single generation each year.

Ventral: the underside.

Vestigial: a remaining bit of something once present, not as functional as it once was.

Bibliography

If you have enjoyed this book and wish to know more details about butterflies of the Caribbean and Florida, some other relevant books are listed here:

Barcant, M. (1970). *Butterflies of Trinidad and Tobago*. London: Collins (out of print).

Brown, M. and Heineman, B. (1972). *Jamaica and its Butterflies*. London: E. W. Classey (out of print).

Gruner, L. and Riom, J. (1979). *Butterflies and Insects of the Caribbean*. Tahiti: Les Editions du Pacifique (translated from French and focused on Guadeloupe and Martinique).

Gerberg, E. and Arnett, R. H. Jr (1989). *Florida Butterflies*. Baltimore, Maryland: Natural Science Publications.

Minno, M. and Thomas, E. (1993). *Butterflies of the Florida Keys*. Gainesville, Florida: Scientific Publishers.

Riley, N. D. (1975). *A Field Guide to the Butterflies of the West Indies*. London: Collins.

Smith, D., Miller, L. and Miller J. (1994). *The Butterflies of the West Indies and South Florida*. Oxford: Oxford University Press (the definitive work on West Indian butterflies).

Stiling, P. (1986). *Butterflies and Other Insects of the Eastern Caribbean*. Basingstoke: Macmillan.

Stiling, P. (1989). *Florida's Butterflies and Other Insects*. Sarasota, Florida: Pineapple Press.

Keen lepidopterists may care to subscribe to the following journals, some of which may be found in your local university library:

Bulletin of the Allyn Museum
Journal of the Lepidopterists Society
Tropical Lepidoptera
Atala
Journal of Research on the Lepidoptera

Geographical distribution

Common name	Scientific name	Florida	Bahamas	Cuba	Hispaniola	Jamaica	Puerto Rico	Leeward Isles	Windward Isles	Trinidad
Danaidae										
Monarch	*Danaus plexippus*	●	●	●	●	●	●	●	●	●
Queen	*Danaus gilippus*	●	●	●	●	●	●	●	●	●
Jamaican Monarch	*Danaus cleophile*				●	●				
Soldier	*Danaus eresimus*	●		●	●	●	●		●	●
Large Tiger	*Lycorea cleobaea*			●	●	●	●		●	●
Ithomiidae										
Jamaican Clearwing	*Creta diaphana*				●	●				
Cuba Clearwing	*Creta cubana*			●						
Blue Transparent	*Ithomia pellucida*									●
Satyridae										
Ringlet	*Calisto sp.*			●	●	●	●			
Night	*Taygetis echo*									
Wood Nymph	*Cercyonis pegala*	●								●
Lady Slipper	*Pierella hyalinus*									●

110

Nymphalidae

Common name	Scientific name	Florida	Bahamas	Cuba	Hispaniola	Jamaica	Puerto Rico	Leeward Isles	Windward Isles	Trinidad
Silverking	*Archaeoprepon demophoon*			●	●	●				●
Ruddy Daggerwing	*Marpesia petreus*	●		●	●	●	●	●	●	●
Cuban Daggerwing	*Marpesia eleuchea*		●	●	●	●				
Common Daggertail	*Marpesia chiron*			●	●	●	●			●
Mosaic	*Colobura dirce*			●	●	●	●			●
Cadmus	*Historis acheronta*			●	●	●	●			●
Orion	*Historis odius*			●	●	●	●		●	●
Cracker	*Hamadryas feronia*			●						●
Queen Cracker	*Hamadryas arethusa*			●	●					●
Jamaican Mestra	*Mestra dorcas*			●	●		●		●	●
St Lucia Mestra	*Mestra cana*			●	●	●	●			●
Florida Purplewing	*Eunica tatila*	●	●	●	●					●
Dingy Purplewing	*Eunica monima*	●	●	●	●	●	●			
Jamaican Admiral	*Adelpha abyla*					●				
Haitian Admiral	*Adelpha gelania*				●		●			
Cuban Admiral	*Adelpha iphicla*			●						●
Trinidad Admiral	*Adelpha cytherea*									●
Mimic	*Hypolimnas misippus*			●	●	●	●	●	●	●
Buckeye	*Junonia coenia*	●	●	●						
Caribbean Buckeye	*Junonia evarete*	●	●	●	●	●	●	●	●	●
White Peacock	*Anartia jatrophae*	●	●	●	●	●	●	●	●	●

111

Common name	Scientific name	Trinidad	Windward Isles	Leeward Isles	Puerto Rico	Jamaica	Hispaniola	Cuba	Bahamas	Florida
Red Anartia	*Anartia amathea*	●	●							●
Huebner's Anartia	*Anartia chrysopelea*							●	●	
Red Rim	*Biblis hyperia*	●	●		●	●	●	●		●
Malacahite	*Siproeta stelenes*	●	●		●	●	●	●		●
Pearl Crescent	*Phyciodes tharos*							●		●
Painted Lady	*Vanessa cardui*									●
American Painted Lady	*Vanessa virginiensis*						●	●		●
Red Admiral	*Vanessa atalanta*				●	●	●	●	●	●
Mexican Fritillary	*Euptoieta hegesia*				●	●	●	●		●
Variegated Fritillary	*Euptoieta claudia*				●	●	●	●	●	●
'89'	*Callicore aurelia*									
Viceroy	*Basilarchia archippus*									●
Red-spotted Purple	*Basilarchia astyanax*									●
Heliconiidae										
Zebra Longwing	*Heliconius charitonius*	●	●	●	●	●	●	●	●	●
Flambeau	*Dryas iulia*	●	●	●	●	●	●	●	●	●
Silverspot	*Dione juno*	●	●	●	●	●	●	●	●	●
Gulf Fritillary	*Agraulis vanillae*			●	●	●	●	●	●	●
Doris	*Heliconius doris*	●	●							
Small Blue Grecian	*Heliconius sara*	●	●							
Postman	*Heliconius melpomene*	●	●							

Trinidad
Windward Isles
Leeward Isles
Puerto Rico
Jamaica
Hispaniola
Cuba
Bahamas
Florida

Common name	Scientific name	Florida	Bahamas	Cuba	Hispaniola	Jamaica	Puerto Rico	Leeward Isles	Windward Isles	Trinidad
Riodinidae										
Bee	*Cholinea faunus*									●
Lycaenidae										
Atala	*Eumaeus atala*	●	●	●						
Giant Hairstreak	*Pseudolycaena marsyas*	●	●						●	●
Grey Hairstreak	*Strynon melinus*	●								
Red-banded Hairstreak	*Calycopis cecrops*	●								
Cassius Blue	*Leptotes cassius*	●	●	●	●	●	●	●	●	●
Hanno Blue	*Hemiargus hanno*	●	●	●	●	●	●	●	●	●
Pieridae										
Great Southern White	*Ascia monuste*	●	●	●	●	●	●	●	●	●
Florida White	*Appias drusilla*	●	●	●	●	●	●	●	●	●
D'Almeida's White	*Appias punctifera*			●	●			●		
Cabbage White	*Pieris rapae*	●								
Little Sulfur	*Eurema lisa*	●	●	●	●	●	●	●	●	●
Sleepy Orange	*Eurema nicippe*	●								
Dwarf Yellow	*Nathalis iole*	●								
Red-splashed Sulfur	*Phoebis avellaneda*		●	●	●		●			
Orange-barred Sulfur	*Phoebis philea*		●		●		●			
Apricot Sulfur	*Phoebis argante*	●	●		●				●	●
Large Orange Sulfur	*Phoebis agarithe*	●	●						●	

Common name	Scientific name	Florida	Bahamas	Cuba	Hispaniola	Jamaica	Puerto Rico	Leeward Isles	Windward Isles	Trinidad
Cloudless Sulfur	*Phoebis sennae*	✓	✓	✓	✓	✓	✓	✓	✓	✓
Orange Sulfur	*Colias eurytheme*	✓								
Dogface	*Colias cesonia*	✓								
Papilionidae										
Cuban Kite Swallowtail	*Protesilaus celadon*			✓						
Haitian Kite	*Protesilaus zonarius*				✓					
Jamaican Kite	*Protesilaus marcellinus*					✓				
Zebra Swallowtail	*Protesilaus marcellus*	✓								
Cattle Heart	*Parides anchises*									✓
Spear-winged Cattle Heart	*Parides neophilus*							✓		✓
Gundlach's Swallowtail	*Parides gundlachianus*			✓						
Pipevine Swallowtail	*Battus philenor*	✓		✓						
Gold Rim	*Battus polydamas*	✓	✓	✓	✓	✓	✓	✓	✓	✓
Devillier's Swallowtail	*Battus devilliers*		✓	✓						
Palamedes Swallowtail	*Pterourus palamedes*	✓								
Green Clouded Swallowtail	*Pterourus troilus*	✓								
Tiger Swallowtail	*Pterourus glaucus*	✓								
Homerus Swallowtail	*Pterourus homerus*					✓				
False Androgeus Swallowtail	*Heraclides thersites*					✓				
Androgeus Swallowtail	*Heraclides androgeus*	✓		✓	✓	✓	✓		✓	✓

Common name	Scientific name	Trinidad	Windward Isles	Leeward Isles	Puerto Rico	Jamaica	Hispaniola	Cuba	Bahamas	Florida
Thoas Swallowtail	*Heraclides thoas*	●				●		●		●
Giant Swallowtail	*Heraclides cresphontes*							●	●	●
Dusky Swallowtail	*Heraclides aristodemus*		●		●	●		●	●	
Bahamian Swallowtail	*Heraclides andraemon*					●	●	●		
Machaonides Swallowtail	*Heraclides machaonides*						●			
Scarce Haitian Swallowtail	*Heraclides aristor*						●			
Cuban Black Swallowtail	*Heraclides oxynius*							●		
Eastern Black Swallowtail	*Papilio polyxenes*								●	●
Hesperidae										
Long-tailed Skipper	*Urbanus proteus*	●	●	●	●	●	●	●	●	●
Sickle-winged Skipper	*Achlyodes mithridates*	●	●	●	●	●	●	●	●	●
Brazilian Skipper	*Calpodes ethlius*			●	●	●	●	●	●	
Brassolidae										
Owl	*Caligo teucer*	●								
Morphidae										
Morpho	*Morpho peleides*	●								

Index

Numbers in *italics* indicate Figures; those in **bold** indicate Tables.